SURVEYING AND MAPPING FOR FIELD SCIENTISTS

WILLIAM RITCHIE
MICHAEL WOOD
ROBERT WRIGHT
University of Aberdeen

DAVID TAIT
University of Glasgow

Longman
Scientific &
Technical

Copublished in the United States with
John Wiley & Sons, Inc., New York

Longman Scientific & Technical,
Longman Group UK Limited,
Longman House, Burnt Mill, Harlow,
Essex CM20 2JE, England
and Associated Companies throughout the world.

Copublished in the United State with
John Wiley & Sons Inc., 605 Third Avenue, New York, NY 10158

This is a new edition of *Mapping for Field Scientists*, first
published by David & Charles Publishers plc 1977
This edition first published 1988
Reprinted 1991

British Library Cataloguing in Publication Data

Surveying and mapping for field scientists.
  — New ed.
  1. Cartography
  I. Ritchie, W. II. Ritchie, W. : Mapping
  for field scientists
  526   GA105.3

ISBN 0-582-30086-X

Library of Congress Cataloging-in-Publication Data

Surveying and mapping for field scientists.

  "This is a new edition of Mapping for field
scientists, first published by David & Charles
Publishers, 1977" — Verso t.p.
  Bibliography: p.
  Includes index.
  1. Topographical drawing. 2. Surveys — Plotting.
3. Aerial photogrammetry. I. Ritchie, William,
1940– . II. Mapping for field scientists.
TA616.S87   1987   526.8   87-3185
ISBN 0-470-20846-5 (USA only).

Set in 9½/12pt Linotron 202 Times Roman

Produced by Longman Singapore Publishers (Pte) Ltd
Printed in Singapore

CONTENTS

Contents

PREFACE

Aims and objectives

This book is designed as a comprehensive practical introduction and guide to a range of sources and techniques that can be used by a wide group of people (collectively embraced by the term field scientists) who are required from time to time to obtain measurements of distances, heights, distributions and dimensions of areas and features in the field. Three main conventional techniques are described and explained.

1. Consultation of existing sources of maps, aerial photographs, etc.
2. Ground surveying.
3. Measurements on aerial photographs and remotely sensed imagery.

These three methods form the core of the book and are followed by a substantial section on cartographic presentation. At the outset, however, there is a short section that introduces the factors involved in the selection of the optimum solution to the particular mapping task in hand, bearing in mind equipment available, personnel, time, cost and, most important, the nature of the end product.

In each section the emphasis is placed on techniques that require basic equipment and resources. In the cartography section, for example, there is little discussion of digital cartography and computer graphics.

In the surveying and photogrammetric sections, more advanced techniques are omitted if it appeared unreasonable to expect that access to the necessary equipment and materials would be available. There is, however, no hard boundary between traditional and new technology and techniques. In the last decade, the advent of cheap hand calculators, and less expensive but increasingly powerful microprocessors, has initiated major changes in mapping science. Electromagnetic distance measurement equipment, once only described in advanced surveying courses, is now seen along many roadsides and in many building sites. Many popular books and magazines contain superb colour photographs from satellites. Accordingly, towards the end of each section of the book, more modern and expensive methods are introduced. Their introduction can be justified not only by an awareness of the rapidity with which mapping science is developing but also in the knowledge that many field scientists may be in a position to commission relatively large and complex mapping programmes. Thus, although the objective of the book is to serve the needs of the field scientist faced with the problem of producing his own map or series of measurements, it is essential to go a little way beyond this basic practical level. In addition, since the emphasis is on empirical solutions, mathematical proofs and theories are, more or less, omitted, but considerable emphasis is placed on recording and translating field and office measurements into the final form of presentation. Indeed, as will be described in the ensuing section, one of the underlying themes of this book is to advocate an approach that begins with a clear definition of the end product and then works backwards to a consideration of the techniques and methods that are available for the achievement of the necessary levels of scale and accuracy. Finally, although the book has to be divided into a series of chapters, it is realised that in many situations there is the possibility that the optimum solution is frequently obtained by combining techniques and sources. This might be illustrated by a simple example of a vegetation survey, whereby the actual distributions are recognised and defined on an aerial photograph which is then taken into the field where the boundaries are confirmed and the composition established. These boundaries are then transferred to an existing base map by some conventional ground-surveying technique.

The main factors affecting the appreciation of the nature of the mapping problem and guidance on how to select the appropriate techniques are given in the subsequent section.

ACKNOWLEDGEMENTS

We are indebted to the following for permission to reproduce copyright material:

British Aerospace for fig. 4.2; Jet Propulsion Lab., California for fig. 4.11; Letraset for fig. 5.8; Macaulay Institute for fig. 4.4 (plate 2); National Remote Sensing Centre, Farnborough for figs. 4.19 & 4.21; Natural Environment Research Council for fig. 4.10; Royal Aircraft Establishment, Space Dept. for fig. 4.7; Spectral Data Corporation for fig. 4.20; University of Michigan for fig. 4.14; Westinghouse Corporation for fig. 4.8

Chapter 1 THE PROBLEM, AND HOW TO APPROACH A SOLUTION

1.1 MAIN FACTORS AFFECTING CHOICE OF TECHNIQUE

Faced with the problem of producing a map which determines a distance or an area, or demarcates a zone of interest, the field scientist may consider the tasks of topographic surveying or mapping from aerial photographs to be either outside his training and knowledge, or routine tasks to be completed roughly and quickly (and consequently inaccurately) by the first means available. Similarly, the use of remotely sensed sources is either a most superficial exercise, or is perceived as a daunting technical task. Neither of these two extreme viewpoints can be justified. As demonstrated in subsequent sections, there is always at least one course of action that realises the objectives of the survey or interpretation, conveniently and economically. Basic topographic surveying or elementary photogrammetry contain no techniques or principles that are beyond the ability of anyone with basic mathematical and fieldwork capacity. Common sense is more important than mathematical or technical ability. The initial perceived difficulties can be resolved if a few basic rules are satisfied and if there is a willingness to consider alternative solutions to the particular problem. Perhaps the most important 'rule' is for the surveyor or photogrammetrist to ask the questions, 'What is the nature of the end product?' and 'What are its format, content and accuracy requirements?'

One of the most common procedural errors is a loose definition of the real purpose of the survey. Three elements must be considered. What are the boundaries of the area of survey? What are the dimensions of the smallest object or distance that must be shown clearly on the final map or diagram? (This concept might be termed 'map resolution'.) Will measurements of areas or distance be taken subsequently from the final map? The answers to these questions determine the scale of the final graphical output, be it a map or a diagram. The surveyor must remember that distances, areas and dimensions can only be plotted (or measured) with an accuracy equal to the finite limits imposed by the scale factor, line thickness and the plotting or measuring device used, e.g. a ruler.

An appreciation of the limitations imposed by the scale of this end product, be it diagram or topographic map, is one of the fundamentals of efficient surveying. If, for example, the finest line plotted on the final map is 0.2 mm, then, at the large scale of 1:1000, on the map it is equivalent to 1000×0.2 mm, or 0.2 m in the field. At the reduced scale of 1:10 000 this line becomes 2.0 m on the ground (see Table 1.1). The knowledge that the scale of the final map determines the limits to which one requires to measure objects or distances has several practical consequences. The field equipment or, if one is using aerial photographs, the measuring devices used, should always permit measurements to be made to an accuracy greater than the predetermined plottable error. As a rule-of-thumb, a field accuracy of about half the plottable error is normal.

The constraints imposed by scale do not apply if the end product of the fieldwork or photographic measurements is expressed numerically. The physical limitations of line thickness, paper size and scale requirements are irrelevant: it is the inherent precision of the equipment used and the skill of the surveyor that are important. Whether a distance is measured as 102.7 m or 102.74 m depends on the capacity of the equipment used and the precision of its manufacture. The skill and training of the surveyor determine whether or not any specific item of equipment is used to its full accuracy potential. The fallibility of human recording of observations is well known, and most surveying practices

1

include checks and other procedures that reduce the chance of errors and, at worst, indicate that an error has been recorded somewhere in the survey work. Simple rules of procedure such as 'working from the whole of the part' also confine errors and discrepancies to well-defined limits or controls.

Once the surveying problem is defined and the decision taken on the manner and, if applicable, scale of presentation, the following factors should be considered.

1. Provision of existing surveyed information and data (usually in map form but including aerial photographs, and lists of coordinates/heights).
2. Provision of control.
3. Time available.
4. Cost.
5. Equipment available.
6. Training and skill of survey team.
7. Number of assistants available.
8. The logistic situation.

Quite often, expensive and laborious topographic surveys are made when adequate maps already exist. Ignorance of the existence of a wide range of maps and other published data at various scales for many parts of the world is widespread, particularly among those whose normal vocation does not require them to use maps other than road maps or popular series. The ensuing section (1.2) on source evaluation outlines, in particular, the major existing series of maps of Britain and indicates availability and coverage for other regions. The first question for a surveyor is, therefore, 'What maps exist for this area?' A few hours in a map library, local planning office or regional office of the Ordnance Survey may obviate the need for many weeks of field work.

Ignorance of the existence of aerial photographs is even more widespread; there are few parts of Britain, for example, that do not have reasonable aerial photographic coverage which is at least adequate for the reconnaissance stage of every survey project. Additional questions may then be asked: 'Is field mapping necessary?'; or 'Does the aerial photograph display the desired distribution?' It is thus possible that the best way of conducting a topographic survey is to use aerial photographs or other remote sensing techniques. A combination of field and aerial-photographic measurements may be the ideal solution to a given problem.

An existing map will rarely fulfil exactly the requirements of the field scientist, but it may be taken as a base and the worker's own surveyed data can then be added to its outline. If aerial photographs are used, this work becomes a process of transference and several simple techniques are available (see Ch. 3). If field methods are preferred, the base map may be taken into the field and information and measurements added to it (see Ch. 2), or the additions may be compiled later in an office or laboratory with or without an intermediate computation stage.

Control is almost synonymous with 'framework' or 'limits', and refers to the skeleton around which the body of the survey is built. This is discussed in Chapter 2, but, as a simple example, consider the planimetric survey of a quadrilateral area of ground whose four corners, A, B, C and D, form the control framework (Fig. 1.1). The corners determine the limits of the area to be surveyed. Moreover, since topographic detail or the dimensions or locations of features are enclosed within this framework, its accuracy of construction defines the limiting (best) accuracy of the total survey. It therefore follows that the measurement or procurement of this framework should be made to an order of accuracy that is higher than the specifications of the topographic mapping. A distinction may be drawn between control measurements, which may be used in computation and subsequently to produce other measurements, and the measurements made to obtain the topographic detail that is surveyed within the control framework. The survey of detail is often plotted to graphical accuracy, and it can therefore be made to a lower order of accuracy than the framework. Equivalent control procedures apply to the use of aerial photographs for surveying purposes.

The control framework may be determined by

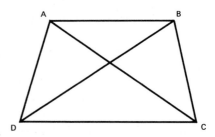

Fig. 1.1 Braced quadrilateral control framework.

one of the following methods: ground surveying; abstraction from aerial photographs, existing maps or charts; or the acquisition of numerical values. The coordinates of the corners of the quadrilateral may be used to determine this control framework, or, for heighting, the Ordnance Survey (and equivalent national surveying organisations elsewhere in the world) provides lists of bench-mark elevations to which one can relate the height information of the particular topographic survey.

It is worth noting that the framework lines are often omitted from the final map, since the purpose of the map is usually to present details of the ground within the area of the control. The control framework, therefore, may be compared to scaffolding – necessary for construction but removed in order that the final product can be seen.

Surveying is expensive, and the cost rises rapidly with area, accuracy requirements and amount of detail surveyed. One of the major cost components is related to time. Preparation work, provision of materials, survey measurements, computations and draughting can absorb many days of skilled work. It is highly pertinent to see the solution to the problem of surveying an area as a time/cost/accuracy equation. This approach places a premium on selecting the right method for each task, reducing measurements to the minimum, eliminating time-wasting errors and using the equipment that most rapidly satisfies the precision requirements. Measurements should not be taken to millimetres if the plottable error is measured in centimetres. This does not mean that an instrument which measures to millimetres should not be used: it may be quicker to use, although it will probably cost more to buy.

The cost of equipment should be related not only to its precision capabilities and its speed of operation but also to the number of functions it can perform. A theodolite used as a tacheometer can measure distance to an accuracy of 1:500 to 1:1000, but will currently cost over £1500; a man can pace out a distance with some degree of accuracy and the 'equipment' cost is nothing. A theodolite, however, can measure angles in the vertical and horizontal planes, distances, and can indirectly measure heights; a man's pace can only measure distance.

If funds are short and the surveyor is provided with a limited range of equipment from existing stock, then, assuming he has the skill to use the equipment, the choice of method is circumscribed by availability. If there is adequate funding, the costing exercise must extend beyond capital purchasing to an evaluation of operator time, speed, flexibility and precision against the demands of the end product.

In other circumstances, it may be worth considering hiring from one of the survey manufacturers or their agents a piece of equipment for a special task. The possibility of borrowing equipment from an educational institute or the engineer's office of a local authority could also be investigated. Other factors enter into the decision-making process. Is there a sufficient level of skill and training available in the survey party? Is there a sufficient number of assistants available to do ancillary tasks such as staff-holding or booking? Is logistic support required for the transport of equipment and personnel?

Any field survey can be broken down into several stages: definition of requirements; reconnaissance; appraisal of methods; field work; compilation and computation (if necessary); and presentation. Each stage poses inter-related questions. The evaluation of the total field survey can be likened to a flow diagram where the earlier stage conditions the next one or permits complete sections of the evaluation to be bypassed. Such an approach is shown in Fig. 1.2, which may be used as a check-list before beginning any ground-surveying project. It also serves as a summary of the factors involved in planning a survey and as a memorandum of the alternatives available.

1.2 SOURCES OF MAPS AND AERIAL PHOTOGRAPHS

The surveyor's problem should be defined clearly at the outset, e.g. 'a planimetric map of an area 1 km square containing field boundaries . . . and on which can be represented short walls, 50 cm thick, to scale'. The search for a solution may be brief, since a suitable map may already exist in an official map series or as a commercial product (Fig. 1.3). Photographs may also be available, either for the direct tracing of detail (in certain circumstances

3

Fig. 1.2

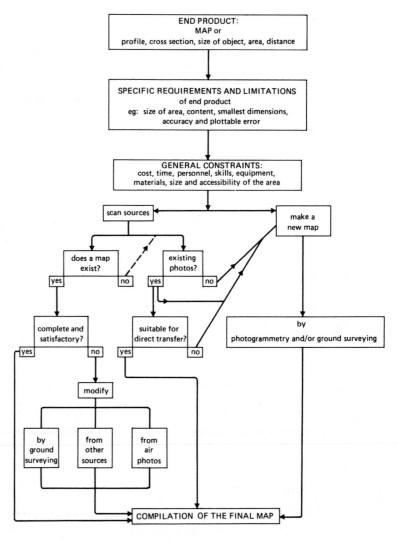

only) or for use directly in mapping by photogrammetry (Ch. 3). The available map may be complete, though it is more likely to require revision, modification or amplification in order to meet the project specifications. This additional work could be completed either by ground survey or from aerial photographs. Possible short-cut solutions, through the use of existing maps or photographs, should be explored at the outset. Most forms of survey should not be proceeded with before a routine examination of existing sources, which may produce a map that requires only a few days' work to bring it to the desired standard. If the area is small and the level of complexity low, however, it is possibly more expedient to make the map directly.

1.2.1 MAPS

Although some surveying techniques may be used merely to add items of information to small-scale maps (e.g. the barometer at 1:50 000), the most common mapping problems assumed for this book would generally relate to small areas (e.g. 100 m square to 2 km square), thus making it unnecessary to consider map sources at scales smaller than 1:25 000. The 1:25 000 scale map, useful in certain situations, has been derived in some regions from larger scales through a degree of generalisation of form and content which is undesirable in maps normally used by field scientists. But in some countries, such as the USA, the corresponding map,

4

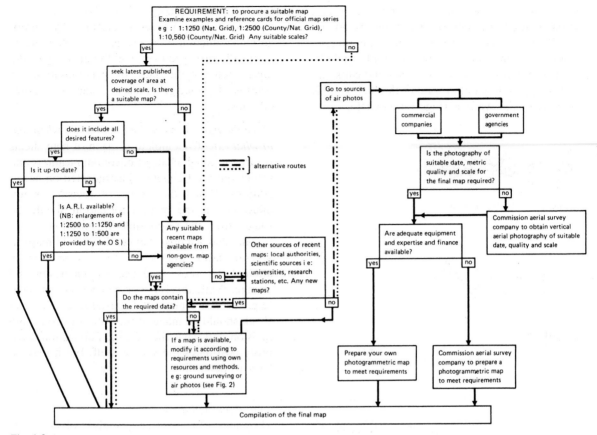

Fig. 1.3

1:24 000, may have been produced to a high order of accuracy, there being no other official larger-scale maps available.

The major map sources useful to surveyors are official maps which are currently published or are of recent date. Other maps may be available from private companies and earlier scientific surveys (Fig. 1.3).

The following questions should be asked when evaluating maps as basic information sources.

1. *What are their scales, their outstanding characteristics of design and content and the areas for which they are available?*

   Current official map series normally cover extensive regions. Small-scale maps (e.g. 1:50 000) are normally available for the whole inhabited area of developed countries, but very large-scale maps (e.g.1:2500) have coverage restricted to well-populated areas on the grounds of economics and requirement. The Ordnance Survey can provide details of the official mapping of

Britain.) In developing countries or sparsely populated areas, complete map coverage, even at small scales, is seldom available.

As the mapping of a country evolves, several editions of one map scale may be available for different regions, and users should anticipate the problems associated with this. It is important to become familiar with the system for locating the sheets in a series, and with the coordinate system used to locate features within any given sheet. These matters are normally explained in the map margin.

2. *How accurate are the maps?*

   Survey work in the nineteenth century was occasionally inconsistent and generally less accurate than similar modern work. Accuracy varied regionally, more remote areas receiving less rigorous attention. These variations are seldom obvious, as mapmakers preferred to give the impression that all areas were treated equally well, and the design echoes this apparent uniformity of standard! For example, in Britain, treatment of trees and shrubs at the 1:1250 and 1:2500 scales is conventional rather than precise, and trees are indicated by different symbols. Other features, such as sand dunes and cliffs, are not mapped in detail in spite of the high graphic quality of the symbols (Figs 1.4 and 1.5).

coastal rocks – older large-scale map series

coastal rocks – air photo interpretation

Fig. 1.4

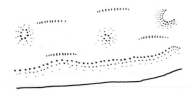

sand dune coast – older series

sand dune coast – from modern photogrammetric map
(contour interval: 4 m)

Fig. 1.5

Attention to notes in the margins of the map sheet on date of survey and compilation is consequently important.

The problem of inconsistency in accuracy and content is exaggerated in smaller-scale maps that have been compiled from larger-scale map sheets of varying quality. In addition to this unapparent variable reliability, the content may have been generalised, be highly subjective and thus unpredictable. Today, certain new maps, e.g.

the Ordnance Survey 1:25 000 Second Series, may have been produced partially by photographic reduction from a larger-scale photogrammetrically surveyed map, and the intricacy of the image reflects the lack of generalisation, in contrast to the generalised image of earlier editions.

3. *Have the maps been revised, and if so, when and to what extent?* Complete revision of map sheets is rare, although changes in communications and settlement are frequently incorporated on newer editions. When the extent of revision is not indicated on the map, information should be sought from the responsible organisation.

   The field scientist, having discovered maps at several dates and scales for his area, may attempt to seek evidence of environmental change. While this is possible in certain instances (e.g. successive maps of a coastal spit), a careful check should be made of the dates of survey and revision and what they mean. Publication obviously post-dates the survey itself, but it may be hard to determine by how long.

4. *What features are normally included in large-scale topographic maps?*
   The choice of content or symbols in official maps evolved in response to military and civilian needs over the years. The selection of features and the design of certain symbols may date from mid-nineteenth-century surveys, and thus will not conform to current design or user requirements. Consequently, the Ordnance Survey in Britain undertook a re-examination of the content of the 1:1250, 1:2500 and 1:10 000 scale maps. Users should be aware of the origins of the maps they select and the consequent effect on their appearance.

1.2.1.1 OFFICIAL MAPS OF GREAT BRITAIN

Three official organisations produce maps of the British Isles: the Ordnance Survey, for Great Britain; the Ordnance Survey of Northern Ireland; and the Ordnance Survey of the Irish Republic.

Table 1.1. presents the range and some of the characteristics of the larger-scale maps produced by the Ordnance Survey of Great Britain.

Table 1.1 Characteristics and coverage of larger-scale Ordnance Survey maps

Scale	The approximate ground width represented by a map line of 0.2 mm gauge	Coverage of one sheet	National Grid line intervals depicted	Planned area of coverage	Remarks
1:1250	25 cm (10 in. approx.)	500 m × 500 m	100 m	Towns of 20 000+	Completely new, introduced in 1945
1:2500	50 cm (20 in. approx.)				
County Series		1½ ml × 1 ml	None	As above and rural areas	Adopted in 1840s
National Grid Series		2 km × 1 km	100 m	As above	Introduced in 1949, but production halted in 1973 in areas where 1:1250 existed
1:10 560	2 m (6.5 ft approx.)				
County Series		6 ml × 4 ml (full sheet) 3 ml × 2 ml (quarter sheet)	None, although some received National Grid	Whole country	Introduced in mid-nineteenth century; quarter sheet came later
National Grid Series		5 km × 5 km	1 km	As above	Appeared after 1945 and includes contours at 25 ft intervals which vary in accuracy
1:10 000		As above	As above	As above	Metric specification adopted in 1969 for all new '6 Inch' maps, 5 m vertical contour intervals (10 m in mountains)
1:25 000	5 m (16.5 ft approx.)				
Provisional Edition		10 km × 10 km	1 km	Almost all but Scottish Highlands and Islands	Published after 1945
Second Series 'Pathfinder'		10 km × 20 km	1 km		Commenced in 1965

All new maps are based on the National Grid. This is a system of lines parallel to and at right-angles to the central meridian (2° W) of the Transverse Mercator Projection, on which the maps of Great Britain are drawn. The grid lines are scaled in metres and the resulting rectangular coordinates can be used to make reference to any point in Great Britain. The origin (zero) of the grid system lies on the central meridian, but, in order to provide positive eastward and northward values, a so-called 'false origin' has been selected about 150 km west and 20 km south of Land's End. The precision achieved with such grid references depends upon the scale of the map. An accuracy of 10 m can easily be estimated on the 1:1250 map, while at the 1:50 000 scale this drops to about 100 m. An index of the National Grid squares and their designated letters appears inside the covers of most folded Ordnance Survey maps and in the map catalogue.

The two largest scale maps, printed in black only, contain detailed information of buildings, communications and boundaries, with spot heights but no contours. The smaller scales, 1:10 000 and 1:10 560, have both spot heights and contours, although the latter vary widely in reliability. On these maps contours are shown in brown. The maps also contain important control information for plan and height which can be used directly for some lower-order mapping purposes, e.g. amplification of existing maps by ground survey or photogrammetry. For more precise work the Ordnance Survey publishes sheets containing the descriptions, exact coordinates and current heights of benchmarks (Appendix 1).

The earlier County Series versions of both 1:2500 and 1:10 560 were constructed on Cassini's Projection, centred on local meridians. This projection leads to distortion of angles which increases rapidly away from the selected central meridian. The projection could therefore only be applied to small local areas (groups of counties – hence the name 'County Series'), unlike the Transverse Mercator Projection which is now the basic national projection and serves the whole country from one location of its central meridian. The complex variety of sizes and interlocking regions of the County Series led to the occurrence of unequal degrees of distortion at the common boundary of two or more county groups. Mismatching of detail at map edges is the outcome. Such failings contributed to the eventual demise of the series, but since the maps will remain in use for some years, the Ordnance Survey provides the National Grid coordinates for the corners of 1:2500 County Series sheets to allow comparison with new maps based on the National Grid.

The National Grid Series of the 1:10 560 map appeared first as a provisional edition compiled from the old maps but on National Grid sheet lines. The most recent Regular Edition is being produced from original surveys, where possible, or is derived from the larger scales where they exist. The earliest 1:10 560 map had a fine intricate design, with very small point and line symbols; later these were changed slightly, with the introduction of the Provisional Edition. The very much bolder, coarser, new Regular Edition will not withstand much further enlargement. It was planned as a base for detailed field mapping.

Revision information

Owing to the cost and time taken in production, revised editions of large-scale maps are slow to appear, so the information on the published sheet may be considerably out of date. Once the maps at the 1:1250 and 1:2500 scales have been published, however, local Ordnance Survey surveyors are responsible for carrying out continuous revision of the master survey drawings (MSDs) of their own areas.

This information is available in two forms.

1. SUSI (Supply of unpublished survey information). It is possible to call at local Ordnance Survey Offices to ascertain if the relevant MSDs provide the information required. Copies can also be provided at some offices. This is the most up-to-date form in which local survey data can be obtained.

2. SIM (Survey information on microfilm). Microfilm copies are produced of published maps at 1:1250 and 1:2500; and Master Survey Drawings at fixed stages of revision. Ordnance Survey Microfilm agents in various parts of the country carry the latest issues of these and can provide enlarged printed copies, at the appropriate scales, on film or on paper. Microfilm copies can also be purchased, but full details should be sought from the

Ordnance Survey in Southampton. SUSI and SIM are not available for the 1:10 560 or 1:10 000 scale maps.

Various other services are available for large-scale maps such as enlarging, i.e. 1:1250 to 1:500, 1:2500 to 1:500, 1:2500 to 1:1250 and 1:10 560 to 1:10 000. Reductions will also be prepared from 1:1250 to 1:2500 and 1:10 000 to 1:10 560.

Since the early 1970s the O.S. has been preparing FORTRAN-readable computer tapes containing the large-scale maps, including the 1:10 000, in digital form. This expanding service has the advantages of making it possible to plot selected information for customer selected sheet lines – without having to accept the rigid sheet boundaries of the standard map series; specialised advice can be obtained from the O.S.

The new 1:25 000 series (Pathfinder) provides, for the Scottish Highlands and Islands especially, a new and detailed record of the landscape. It was produced for an area whenever the appropriate sheets at the 1:10 560 or 1:10 000 scales had been completed. The aerial survey method employed for the most recent maps has permitted very detailed depiction of rock exposures. The facility of obtaining gradient measurements from the contours is now greatly improved. Elsewhere, various methods are used for major revision of the older editions. The 1:25 000 does suffer from extensive generalisation, e.g. the exaggeration of road width, the simplification of building outlines, and the broad classification of trees. It is worth observing, however, that unlike the Provisional Edition, the new Regular 1:10 560 and 1:10 000 maps have almost the same level of generalisation as the 1:25 000 scale maps, in spite of the difference in scale.

Users of Ordnance Survey publications should become familiar with all the series with SUSI and SIM, where available and also with the copyright regulations governing their use. The most valuable initial source of information is the current O.S. Catalogue and the Ordnance Survey Publication Reports, which may be obtained from the Southampton headquarters of the Ordnance Survey (Appendix 1). Early editions of County Series maps, if required, are still available in some of the larger libraries, and elsewhere, but the Ordnance Survey will provide advice on this.

1.2.1.2 OTHER SOURCES OF MAPS

Other official mapping is generally of a thematic nature, and details of Soil Survey and Geological Survey publications can be obtained from the Ordnance Survey. Land utilisation maps, however, are obtained through the Geography Department, Kings College, University of London. Important addresses, etc., appear in the current Ordnance Survey map literature. In Britain, private mapping is carried out by several commercial firms (Appendix 2). A more complete list for most countries might be obtained by reference to the corporate membership list of an appropriate professional society; these lists often appear in journals such as *The Survey Review*, *Photogrammetric Record* and *Photogrammetric Engineering and Remote Sensing*.

Commercial firms carry out varied work. In Britain, much of it consists of contouring and updating Ordnance Survey large-scale maps for particular clients, notably local authorities. Original mapping from new, specially commissioned photography is also carried out. These companies do not provide a 'library service' of the work undertaken, for, strictly speaking, it is the private property of the clients. For seekers after source maps, carefully worded enquiries to various organisations may bring favourable results. On the other hand, initial contact with newly surveyed material may be obtained through a local planning office.

Similar source facilities and services exist in many countries (Appendix 2), some details of which follow.

The United States of America
In the USA the largest-scale map series is 1:24 000, and this series, with the other official topographic maps, is the responsibility of the US Geological Survey. The sheet lines of US maps are based on the earth's graticule system of parallels and meridians, the 1:24 000 measuring $7\frac{1}{2}$ minutes of latitude by $7\frac{1}{2}$ minutes of longitude, and the 1:62 500, 15 minutes by 15 minutes.

In general, the content of the 1:24 000 map is similar to that of the Ordnance Survey 1:25 000, but, what is more important is that the topographic detail is frequently much fuller and more precise. Not only are there additional symbols for surface deposits, etc., but the contour interval varies from sheet to sheet according to the nature of the topography. The commonest intervals are 10 ft and 20 ft (the Ordnance Survey standard interval has been 25 ft, although 5 m and 10 m intervals are being used in the Second Series), but in flat land (e.g. coastal), intervals of 5 ft or even 1 ft are employed, with 40 ft in mountains. As these contours are being produced almost entirely by photogrammetry, and accuracy standards claim that at least 90 per cent of elevations interpolated from contour lines shall be correct within one-half the contour interval, these maps are extremely valuable to the field scientist. The standards also refer to the 1:62 500 scale, which commonly carries 20 ft and 40 ft intervals, although 10 ft and 80 ft intervals are also employed. This smaller-scale map, also covering more than half the country, may often, therefore, find considerable application in certain field mapping situations. Coverage also exists, in some areas, at 1:31 680, but this older map scale has been replaced by the 1:24 000 series. Alaska is mapped at 1:63 360 only, with 50 ft or 100 ft contour intervals.

A service similar to SUSI also exists in the USA. One-colour advance prints are available of topographic maps in preparation. These can be provided at various stages of completion: manuscripts compiled from aerial photographs; unedited advance prints (field mapped and checked but with no names); and partially edited prints (finally drawn and including names). Details of this service can be obtained from the USGS and prices are listed in their 'Advance Materials Index'.

Canada

Being largely uninhabited, less than 50 per cent of the Canadian land area has been mapped at other than very small scales. The official 1:50 000 coverage includes all regions affected by man, but the 1:25 000 scale map has a very restricted and patchy distribution, mainly around city areas with a population of over 35 000, with its maximum continuity from northern Lake Erie to Quebec. More than half the planned coverage of the 1:50 000 scale map has appeared, and a proportion is available only as single-colour advance prints. The content and accuracy of these two map series, however, make them of considerable value to certain field scientists with a mapping problem. The 1:50 000 series carries contours at 25 ft, 50 ft or 100 ft intervals, while the 1:25 000 series has intervals of 10 ft or 25 ft, depending on topography. Larger scales are not available. One relieving factor is the existence, for the whole of Canada, of vertical air photography of photogrammetric quality and at various scales. In addition, the Surveys and Mapping Branch in Ottawa has been developing, over the past ten years, an immensely valuable Topographic Digital Data Base, full details of which should be sought before any other research is carried out.

Australia

Australian maps at scales of 1:50 000 and larger are the direct responsibility of the State Lands Departments. The earliest programme, started after the Second World War, was of 1:31 680 scale mapping to aid planning and development, but inevitably this had restricted coverage and gradually went out of date. Mapping proceeded at local and national level, but not until 1958 was a national mapping programme set up to provide, initially, maps at scales of 1:50 000, 1:100 000 and 1:250 000 which replace the imperial scales of 1:31 680, 1:63 360 and 1:253 440. Although the older map scales still provide the only coverage in many areas, some state mapping authorities are forging ahead not only with the 1:50 000 and 1:25 000 scales (for the more densely populated areas) but also with topographic/cadastral maps at 1:10 000 for urban regions. Even larger scales are envisaged, the Central Mapping Authority of New South Wales, for example, having planned 1:4000 and 1:2000 scale orthophoto-mapping of the Newcastle-Sydney-Wollogong zone. The Royal Australian Army Survey Corps has a tradition of topographical mapping and still continues to produce maps at various scales, e.g. 1:12 500 coloured orthophoto maps. The National Mapping Council is the control and liaison body for all Australian mapping, including geology, and initial enquiries could be directed there. However, State departments can provide comprehensive information on map availability for local areas.

New Zealand

Apart from a very limited coverage at 1:25 000, the largest current basic map scale in New Zealand is 1:63 360, commonly with only 100 ft contours (for topography and cadastre). This was started in 1936 and is only now reaching completion. However, it has decreased in value with age and the need for larger scales and more accurate maps hastened the inauguration of a range of new map series, cadastral and topographic, which is now expanding.

The larger-scale new topographic series comprise the following: 1:50 000, produced by photogrammetry with 20 m contour interval; 1:25 000, as compiled for basic maps at 1:50 000, with the same contour interval, etc., but available as plan prints only; and 1:10 000, a new series on a photogrammetric base with a 5 m contour interval, but for urban and developing regions only. These, too, will be available as plan prints.

A parallel new series of cadastral maps is also under way, and scales of 1:2000 and 1:1000 are planned for a new record map system. Naturally, existing series will be maintained until the new map sheets become available. Although field scientists have little to choose from in current map series and thus more often rely on air photography, the future is distinctly brighter. All enquiries should be directed to the Land and Survey Department in Wellington.

The following section contains short notes on the official basic larger-scale topographic maps produced by the national agencies of certain European countries. Readers should make fuller enquiries if they need information on special cadastral, water-control or thematic maps. Most developed countries will also have separate organisations producing larger-scale maps for planning, etc. A general statement of map availability can be found in one of the major map bibliographies (e.g. Parry and Perkins 1987). Details of such sources may also be obtained through the information service of the national topographic agency of the country concerned.

Austria

Austria's 1:25 000 scale map series had to be discontinued in 1959 for reasons of economy when it was still far from complete. There is no revision programme for the sheets that were produced although enlargements of the 1:50 000 series are being printed on request. Thus the 1:50 000 series with a 20 m contour interval is now the largest basic scale and includes hachures and hill shading. Since the demise of the 1:25 000 series, on which it was previously based, all mapping has been done photogrammetrically at 1:10 000 and then reduced.

Belgium

Although Belgium has a map series at 1:10 000, printed in 4 colours, it is not genuinely large scale in origin, but has been produced by enlargement from the main basic scale of 1:25 000. This scale also provides complete coverage, is printed in 6 or 7 colours and has contour intervals of 1, 2 or 5 m depending on the terrain.

Finland

Finland is well served by excellent basic maps at 1:20 000. For the purpose of mapping, the country is divided into two regions, north and south. The more populous south has a detailed 6-colour map series with 2 or 5 m contour intervals. This map is compiled at 1:10 000 and, although not printed at this scale, copies can be obtained if required. The northern area, with its lower population density, also has 1:20 000 map coverage but is printed in only 3 colours and with contour intervals at only 5 or 10 m.

France

The basic topographic series in France has always been at the 1:20 000 scale, although more recently 1:25 000 has been adopted. New sheets are published at this scale and the older 1:20 000 series is progressively being replaced. Contour intervals of 5 m are used in flatter regions with 10 m in mountains. This series is virtually complete for the whole country.

German Federal Republic

The pattern of map coverage in the Federal Republic of Germany is complex and it is difficult to do justice to the details in a short account. Large parts of the country, especially in the south, have a variety of cadastral maps at 1:2500 and 1:50 000 with topographic and relief content. There is also a basic uniform series at 1:5000 which, although not yet complete, will eventually cover 60 per cent of the country. These 2-colour maps have contour intervals of 1, 2.5 or 5 m. The smaller-scale map at 1:25 000 is available for the whole country but, having been produced between 1875 and 1960 and with much of the northern area printed in one colour only, it has certain drawbacks. Nevertheless, the series has been given a major overhaul and is now published mainly in 3 or 4 colours.

German Democratic Republic

This also has a thorough map coverage – all published series being complete. The basic scale of 1:10 000 is printed in 5 colours with a contour interval of 1 or 2.5 m while the 4-colour 1:25 000 series has a 5 m contour interval.

Greece

The largest scale map of Greece, at 1:50 000, is printed in one colour and is almost half complete. It has a contour interval of 4 m. The only other larger-scale maps available are at 1:10 000, but this series was discontinued at a very early stage with only 2 per cent of the planned sheets complete.

Italy

Italy has only one large-scale map series, that of 1:25 000. It is complete, but having had its origins in the late 19th century there have been several changes in its appearance from one colour to the 5-colour edition of today. The contour interval also varies with the edition, being 5 or 25 m. This series is currently being revised.

Luxembourg

There are two complete large-scale map series of Luxembourg, 1:20 000 and 1:25 000. They are very similar in many ways, both having a 5 m contour interval and both appearing in 4 colours.

Netherlands

The basic map of the Netherlands is really at a scale of 1:12 500 but it is not published as such. Instead, it is used to produce the two major basic national series. The larger scale, at 1:10 000 and produced by photographic enlargement, is printed in one colour only. Although very detailed, it has no contours. The 1:25 000 scale map, on the other hand, which appears in at least 2 colours has contours at an interval of 2.5 m. Both these series are being regularly revised.

Norway

Only one basic map series, 1:5000 is produced for Norway (The Economic Map of Norway) and it is still incomplete. Printed in 4 colours it has a 5 m contour interval and, in addition to normal topographic detail, it contains property and soil boundaries. A regular revision programme exists.

Portugal

Portugal has a fully detailed topographic map series at 1:50 000, but the larger-scale maps in progress, at 1:10 000 and 1:5000, are restricted more to urban areas.

Spain

The largest official series is at a scale of 1:50 000. This is a contoured topographic map which also contains information on vegetation and crop details.

Sweden

The two basic map series of Sweden, at scales of 1:10 000 and 1:20 000 are named 'The Economic Map' and are both printed in 4 colours. The 1:10 000 map, now complete, covers about 60 per cent of the country, and is mainly of land of value for farming and forestry. Some recent sheets have a photomap background. The 1:20 000 map will eventually cover 20 per cent of Sweden. A general programme of revision was started in 1974.

Switzerland

The Swiss basic map, at a scale of 1:10 000, is in one colour only. It carries a general contour interval of 10 m, but with 5 m intermediate contours added in flatter areas. The 1:25 000 series, on the other hand, could not be more of a contrast. Printed in 8 colours with a standard 10 m contour interval, it is rich in slope and rock detail, and is enhanced by hill shading.

Basic large-scale mapping coverage of the rest of the world varies greatly. Some countries, such as Malaysia, Lebanon, India and Brazil, do have maps at scales of 1:25 000 or larger but most others must rely on 1:50 000 (e.g. Congo, Ghana, South Africa, Tanzania, Zambia and Saudi Arabia) or smaller.

It should be emphasised that this outline is provided only to give an impression of the extent of official mapping in these countries. Determined field scientists should seek out other more detailed sources before resorting to a new mapping programme. One week of carefully directed enquiries might be worth many weeks of full-time mapping.

1.2.2 AERIAL PHOTOGRAPHY

Unlike the map coverage, which extends across the whole country systematically at a number of scales and is revised periodically, the aerial photographic coverage of Great Britain is generally incomplete at any particular time period, suffers from a mixture of scales, and lacks any systematic programme of rephotographing. A distinction should be made between those photographs taken primarily for topographic mapping – survey photography – and those taken for interpretation purposes only – reconnaissance photography. Both types exist in large numbers, but the reconnaissance variety is generally less suitable for map construction.

1.2.2.1 RECONNAISSANCE PHOTOGRAPHY

Most of this photographic work in Great Britain was done by Royal Air Force units, mainly during the 1940s and 1950s. The photographs are commonly of the 'fan' type, where economic ground coverage is obtained by tilting the cameras away from the vertical. Up to six cameras may operate simultaneously from the same aircraft. In the simplest case, twin cameras are tilted 10° from the vertical to provide what are termed 'split verticals' (Fig. 3.32). The most common photo scale is about 1:10 000, and where high accuracy is not essential the transfer of plan detail from these photographs on to published Ordnance Survey maps at 1:10 000 or 1:10 560 can be effected by means of a Sketchmaster (Fig. 3.31). This RAF photography, generally on a 9 in. by 7½ in. format, is held by the Scottish Development Department (for Scotland), from whom copies may be bought.

1.2.2.2 SURVEY PHOTOGRAPHY

This photographic work has been done by high-quality aerial survey cameras with low distortion lenses, generally on a 9 in. by 9 in. format, with a near-vertical camera axis. The photography can generally be recognised by the presence of characteristic edge markings (such as small crosses in the corners), and the appearance along one edge of

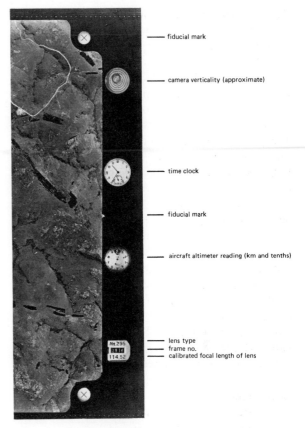

fiducial mark

camera verticality (approximate)

time clock

fiducial mark

aircraft altimeter reading (km and tenths)

lens type
frame no.
calibrated focal length of lens

Fig. 1.6 Marginal information on an aerial photograph.

information about the time and date of photography, a vertical bubble, an altimeter scale, the photograph number and the calibrated focal length of the lens (Fig 1.6).

1.2.2.3 THE ORDNANCE SURVEY

The Ordnance Survey uses survey-quality aerial photography for much of the basic mapping of Great Britain. The main scales of photography available are about 1:23 000, 1:10 000 and 1:7 500. Some areas are photographed at 1:15 000 and some urban areas at 1:3000. Other photographic scales are uncommon. Almost every part of Great Britain has been photographed on near verticals in this manner, and in some areas photographs of several scales and dates are available. The Ordnance Survey has also photographed much of the coastline on black and white infra-red film, mainly for the demarcation of high- and low-water marks.

1.2.2.4 THE COMMERCIAL COMPANIES

The other main sources of survey-quality photographs are the commercial aerial survey companies. The photographs are taken under contract, but if the client agrees, there is generally little difficulty in obtaining copies. There is no general register of each company's holdings, but enquiries about the photographic coverage of individual areas will be answered (Appendix 2). The cost is generally greater than from other sources, but the quality of photography is consistently high. In Great Britain the local authorities are regular clients of the aerial survey companies, some commissioning total photographic coverage of their area at regular intervals (e.g. a complete county coverage during census year is established practice), usually at a scale of 1:10 000 or 1:15 000. The commercial companies also take aerial photographs using different types of emulsion: colour and false-colour.

1.2.2.5 AERIAL PHOTOGRAPHIC ARCHIVES

It is often useful, where change through time is important, to refer to photographic coverage of years gone by. Several centres maintain collections of older aerial photographs, and copies may usually be bought if the negative film is available.

1.2.2.6 THE CENTRAL REGISTERS OF AERIAL PHOTOGRAPHY

In order to collate as much information as possible on the available photographic coverage Central Registers have been established in Great Britain: at the Scottish Development Department in Edinburgh (for Scotland) and at the Welsh Office in Cardiff where a separate register for coverage of Wales is maintained. The Ordnance Survey is now the best source of information on airphoto coverage of England, since the Central Register at the Department of Environment has now been disbanded.

1.2.2.7 CAMBRIDGE COMMITTEE FOR AERIAL PHOTOGRAPHY

The University of Cambridge Aerial Photography Unit holds one of the largest collections of aerial photographs, verticals and obliques, of the British Isles. Most of the sites photographed are of historical or archaeological interest, but there are many others, including a fairly comprehensive coverage of National Nature Reserves. The photography dates back to the 1940s, and although much of the early coverage is oblique photography, on a 5 in. by 5 in. format, the more recent photography includes verticals, some in colour, taken with a high-quality aerial survey camera.

For the field scientist, maps and aerial photographs are basic sources of information. The way they are used in a survey project may range from the provision of control and topographic detail to reconnaissance or background information. Whatever the application, the reader is strongly recommended to consider the foregoing source-evaluation procedures at the beginning of a survey.

1.3 OTHER REMOTELY SENSED DATA

1.3.1 THE UNITED KINGDOM

In the United Kingdom, the most important source of information on the coverage and availability of remotely sensed data (other than aerial photography) is the National Remote Sensing Centre, which was established in 1980 within the Space Department of the Royal Aircraft Establishment at Farnborough. The centre produces a User Guide, updated annually, in which the products and services available from the centre are listed. For the most part, the products are derived from earth-imaging satellites, such as Landsat, Meteosat and Seasat. There has been very limited use of non-photographic methods of remote sensing in Great Britain using aircraft. Satellite data is usually available in photographic form (optical data) or in digital form (computer-compatible tape). The digital product retains much more of the original detail than the optical product and, furthermore, opens up the possibility of data analysis and classification by a computer-based image-analysis system. Such systems are available for hire at the National Remote Sensing Centre. The Centre also acts as National Point of Contact for Earthnet, a system for distribution of information and imagery concerning satellite coverage of Western Europe. The Centre acquires all reasonable quality products of scenes of the British Isles and acts as an information centre for queries concerning satellite coverage of other parts of the world.

1.3.2 SOURCES OF IMAGERY OUTSIDE THE UNITED KINGDOM

In other countries there is great variation in the ways in which information and products from remote sensing missions are collected and disseminated. In some countries such data is regarded as confidential and is not readily available (especially aerial photography), whereas in other countries (such as the USA) information and image products are readily available for both aerial photography and satellite imagery.

If in doubt, a useful first point of contact for any country is the national map-making agency, since they are likely to be principal users of remotely sensed imagery. Often, several government agencies (e.g. Geological Survey, Natural Resources Department, Forestry Department) will have commissioned aerial photography for their own purposes and will maintain a record of their own holdings.

In the USA, for example, air photo archives are maintained (among others) by the US Geological Survey, Soil and Conservation Service, Department of Agriculture, National Ocean Survey and EROS Data Centre. Satellite imagery is generally archived and sold to the public by EOSAT from the EROS Data Centre at Sioux Falls, South Dakota, for all of the USA and other parts of the world which are not part of a regional satellite data collection and distribution network.

A useful guide to imagery sources in North America is *Everyone's Space Handbook – A Photo-imagery Source Manual* by Dick Kroeck (1979). Further information on sources of remotely sensed data is given in Appendix 2.

FURTHER READING

Butterworth, B. A., 1983, *A Directory of UK Map Collections.* The British Cartographic Society Map Curators' Group, Southampton.

Kroek, D., 1979, *Everyone's Space Handbook – A Photo-imagery Source Manual.* Arcata, Pilot Rock, California, USA.

Chapter 2 GROUND SURVEYING TECHNIQUES

2.1 INTRODUCTION TO GROUND SURVEYING

As was explained in Chapter 1, ground surveying is the process by which terrain data is obtained by direct measurement in the field. Unlike map compilation methods based on secondary sources such as existing maps or aerial photographs, ground surveying is a primary method and as such offers many advantages to the field scientist. Whether ground surveying is being carried out by professional surveyors or by field scientists, two basic principles must be understood and rigidly applied.

1. The need to work within a framework.
2. The necessity of independent checks.

The need to work within a framework, sometimes expressed as 'working from the whole to the part', or 'interpolating rather than extrapolating', can be illustrated by a simple example. Suppose that it was necessary to measure the external dimensions of a small regular building. Each length could be measured by using a tape, say 30 m in length, or by using a metre stick. In the first case, only two tape readings would be required for each side, namely the zero mark held at one corner and the reading at the second corner giving the length of that side. If a metre stick was used, many readings would be required as the metre stick was moved metre by metre along the length of the wall.

In the first method, the length is being measured as a unit, to be subdivided later when the doors, windows, etc., are added and therefore any error in this total length would also be subdivided. In the. second case, parts are being measured (each part being 1 m) which have to be summed to give the total dimension, and consequently all the errors in the individual measurements will accumulate to give a much larger error in the total length. Interpolation is therefore better than extrapolation and

this principle should be applied whenever possible.

Arising from this principle is a fundamental need to obtain a strong control framework before detail work is carried out. Indeed, it is useful to think of all survey tasks as having two phases: first, the survey of the network of points which will form the control framework; and, second, the survey of the detail. The first part, called the control survey and producing control points, is extremely important but is often rushed or completely neglected by the field scientists in their enthusiasm to survey the details of particular interest to them. However, without a good, strong framework no detail survey will be successful, no resurvey work for checking or change detection will be possible and the location of errors will be much more difficult. The types of framework available are outlined in sect. 2.2.1.6 and are illustrated in Fig. 2.13. The second phase, in which the detail work is carried out, can usually be executed to a lower order of accuracy.

The accuracy of a single measurement can never be guaranteed. One check is to take the measurement again, perhaps by a different method or using a different observer, but even in these cases errors can escape notice. The best type of check is one that is independent of the measurement. For example, if two angles in a triangle are measured, the third angle can be found by subtraction from 180°. By measuring the third angle, an independent check is available. It should be remembered that these checks are being applied to ensure that the survey meets specifications and not so that the highest order of accuracy· is achieved. A triangle summing to 179° might be quite acceptable for a particular survey.

If it is impossible to check a series of observations by independent means, the observations should be taken more than once. By taking the mean of a series of observations of the same quantity, a better estimate of the true value can be obtained, gross errors can be detected and eliminated and the effect of small random errors tends to be reduced.

Field survey work can be carried out to provide: (1) planimetric information only; (2) height and slope information only; and (3) combined planimetric and height information. This part of the book is arranged according to these aims. In each section, a variety of methods is given for the acquisition of the required terrain data with details of the necessary equipment. The main subdivision of this part of the book is therefore according to methods: it is not intended to give a systematic account of equipment. Readers wishing to concentrate on particular items of equipment should refer to the index.

It will be assumed that the maps being produced will be drawn by hand, with a plotting accuracy of 0.2 mm (see Ch. 4).

2.2 HOW TO OBTAIN PLANIMETRIC INFORMATION

Introduction

There are many methods of determining the planimetric position of points, and each technique can provide a range of accuracies, depending on the instruments used and the skill with which they are operated. The first part of this section deals with the methods of determining position; the equipment available is then introduced so that consideration of the procedures and accuracies may follow, thus allowing the reader to select the method most appropriate to his task.

What is meant by the planimetric position of a point? One is obviously referring to a point's location, but the terms position and location only have meaning in a mapping context if they are defined with respect to a fixed reference system. For example, if it is known that point A is 200 m from the point B, one can only state that the distance between these points is 200 m. If a third point C lies 300 m from point A, one can only state that the distance AC is 300 m. Nothing can be said about the position of B with respect to C and points B and C need not lie on the same straight line

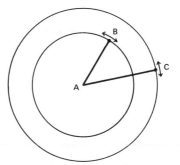

Knowing distances AB and AC does not fix point B and C relative to one another

Fig. 2.1

through A. This is illustrated in Fig. 2.1, where points B and C lie somewhere on circles centred on A with radii of 200 m and 300 m.

Knowing the distance between two points is thus not sufficient for defining a unique location in two-dimensional space. A second item of data is required and that concerns direction. If it is known that point B lies 200 m north-east of point A, and C lies 300 m east of A, the points B and C are now uniquely defined with respect to A (Fig. 2.2). Furthermore, the position of B relative to C is also defined. This system of reference is known as the polar coordinate system, as it depends on bearing (which is defined as the angle measured clockwise from north) and distance from a point of origin. The two parameters required to give a unique location are the bearing measured at the point of origin and the distance from the point of origin.

Point B can be defined with respect to point A using distance measurement only, provided that the distances are well chosen. If the north-to-south and

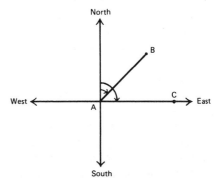

Fig. 2.2 Polar coordinate system.

east-to-west axes of Fig. 2.2 are retained and given a common scale numbered towards the north and east from A (Fig. 2.3), measurement of the easterly distance of B from the north–south axis through A gives the value ΔE, and the corresponding distance measured to B from the east–west axis through A gives ΔN.

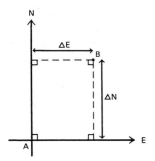

Fig. 2.3   Rectangular coordinate system.

With these two values, B can be fixed with respect to A. This is the system normally employed in graphs and is called the rectangular coordinate system, as position is defined by two measurements in two directions at right angles to one another. Figure 2.4 shows how a rectangular (or Cartesian) coordinate system can be used to plot the position of points whose rectangular coordinates are known.

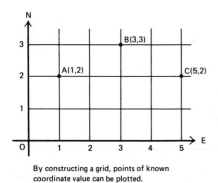

By constructing a grid, points of known coordinate value can be plotted.

For example: A(1,2), B(3,3), C(5,2)

Fig. 2.4   Use of the rectangular coordinate system.

What is the relationship between these two systems of reference? In the polar system, B is defined by the bearing α and the distance D, while in the rectangular system the displacements ΔE and ΔN are used (Fig. 2.5). Hence, from the right-angled triangle ACB,

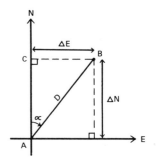

Fig. 2.5   Relation between rectangular and polar coordinate systems.

$$\frac{\Delta E}{D} = \sin \alpha \text{ or } \Delta E = D \sin \alpha$$

$$\frac{\Delta N}{D} = \cos \alpha \text{ or } \Delta N = D \cos \alpha$$

$$\text{and } \frac{\Delta E}{\Delta N} = \tan \alpha$$

Using these simple relations, it is possible to convert from one coordinate system to the other.

Whether the position of a point with respect to a coordinate system is to be determined graphically or numerically, the same rules apply when considering the number of observations necessary to fix or establish the position of a point. In general, at least two observations will be necessary.

2.2.1 METHODS OF FIXING

If there are two points, A and B, of known position within some coordinate system, the distance AB and the bearing of line AB will be known (see Fig. 2.5 and equations above). Methods of fixing the position of an unknown point C will now be discussed.

2.2.1.1 TRIANGULATION

Triangulation is a survey method based on the measurement of angles and it has been extensively used in providing control networks in national surveys. In Fig. 2.6, points A and B are known, while point C has to be determined. If the angles at A and B are measured, the triangle is 'solved'.

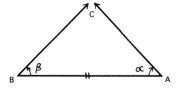

Fig. 2.6   Triangulation.

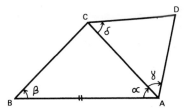

Fig. 2.7

This can be shown graphically in the following way (Fig. 2.7).

With points A and B plotted on a map sheet, a protractor is used to set off or measure out angles α and β. The position of C is at the intersection of the two lines. This procedure of intersection could then be repeated from line AC, which is now determined and by measuring and setting off angles γ and δ, the position of point D could be established.

2.2.1.2 TRILATERATION

While triangulation depends on the measurement of angles, trilateration employs the measurement of distance. If the baseline AB is known, the remaining two sides of the triangle can be found by measuring the two distances AC and BC and the triangle solved (Fig. 2.8). The situation is shown graphically in Fig 2.9.

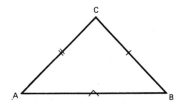

Fig. 2.8   Trilateration.

From A, an arc of a circle with radius AC is drawn. Point C must lie on this arc, as was explained in the introduction to coordinate systems (Fig. 2.1). Simi-

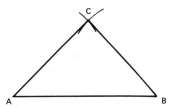

Fig. 2.9

larly, point C must lie on the circumference of the circle centred on B with radius BC. The plotted position of C is found at the intersection of the two arcs.

As with triangulation, AC could now be used as a baseline to extend the trilateration to fix point D, and others.

2.2.1.3 RADIATION

Radiation is a method of fixing which employs both angle and distance measurement and is based on the polar coordinate system. Using the direction or bearing of AB as orientation, the angle and distance to point C_1 are measured (Fig. 2.10). This is followed by other angle/distance pairs to other points to be fixed, e.g. α_2/d_2 to point C_2.

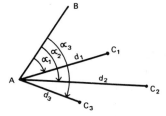

Fig. 2.10   Radiation.

The plotting of these points can be carried out by protractor and scale. In each case, the angle from the orienting line AB is set off using the protractor, thus giving the direction to the unknown point on the plot. The position of the point is obtained by scaling off the required distance along the plotted line.

2.2.1.4 TRAVERSING

Traversing also uses both angle and distance measurement and is very similar in principle to

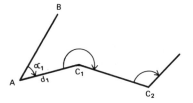

Fig. 2.11   Traversing.

radiation. Starting from a known point A and a known bearing AB (Fig. 2.11), the angle and distance to C_1 are measured as in radiation. The angle to be measured is always the clockwise angle from the old station to the new station. This fixes the position of C_1. If an angle and distance at C_1 to C_2 are measured, the position of C_2 is established. This procedure can be repeated along the length of the traverse.

2.2.1.5 OFFSET OR RECTANGULAR METHOD

This method is based on the measurement of distance within a system of right angles (i.e. the two distances to be measured are at right angles to one another). This method bears the same relationship to the rectangular system of coordinates as does radiation to the polar system.

  With a line AB known, an unknown point C_1 can be fixed by constructing a perpendicular from C_1 to this line and measuring the length of this perpendicular distance and the distance from A or B to the foot of the perpendicular. These perpendiculars are called offsets and they can be measured to the left or right of the main line (Fig. 2.12).

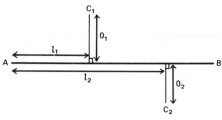

Fig. 2.12   Offset measurement.

2.2.1.6 APPLICATIONS

These five fixing methods are the basis for all planimetric surveying and in survey practice they can

be used separately or in combination. The accuracy obtained using each method will depend on the equipment used for measurement and on whether graphical or numerical methods are employed to determine the position of the new points. These techniques can be used for control or detail surveys, but the accuracy of detail over a large area will be closely related to the accuracy of the controlling framework. In radiation from a single point, this point plus an orienting ray can provide the framework (Fig. 2.13a); in the methods of offset, triangulation and trilateration, a single line can serve as the framework of the survey (Fig. 2.13b); in more extensive surveys, a system of triangles (triangulation or trilateration) or a system of lines (traversing) provide the framework (Fig. 2.13c).

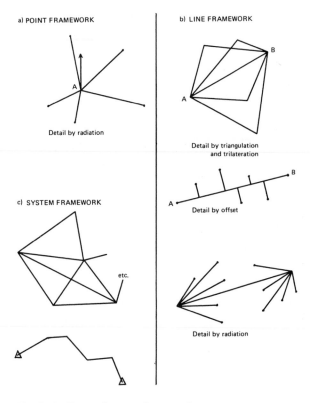

Fig. 2.13   Types of survey framework.

  The most common combinations of method and equipment suitable for use by field scientists will now be discussed, with brief reference being made to other possibilities.

2.2.2. TRILATERATION AND OFFSET BY CHAIN AND/OR TAPE

When the survey area is relatively small, of the order of 0.25 square kilometres, a useful method of obtaining planimetric information is by trilateration and offset. The method is simple to operate in the field and the equipment required is robust and inexpensive.

The two main items of equipment required are the surveyor's chain and the surveyor's tape. There are three main types of chain: a Gunter's chain; a foot chain; and a metric chain. A Gunter's chain is 66 ft in length and is divided into 100 parts known as links. With a chainlength of 66 ft, 10 square chains equal 1 acre, which, in the past, made this chain particularly suitable for measuring lines to be used in the determination of areas when the acre was the standard unit for area measurement. With the almost universal adoption of the metric system of measurement, this advantage is now of little significance (although these chains are still handy for the laying out of cricket pitches). The 100-foot chain and the metric chain (either 20 or 30 m) are now more common, the length of a link in these chains being 1 ft and 200 mm, respectively. Chains are constructed from thick iron or steel wire, each main section being connected by three small, oval rings, the centre of the middle ring marking the end of the link.

A surveyor's tape is usually 100 ft or 30 m in length. Such tapes were once made of linen, but now plastic and plastic-coated metal tapes are more common. Modern tapes are less likely to stretch, are less sensitive to damp and, with care, can be used in place of a chain in many instances. Tapes and steel bands are normally wound on to a spindle in a leather or plastic case and are graduated in feet and inches, feet and tenths of a foot, or in metres and centimetres. Care is needed to ensure that the corrct zero mark on the tape is used (Fig. 2.14).

Other necessary items of equipment are metal arrows, ranging poles and wooden pegs (Fig. 2.15).

2.2.2.1 USE OF THE CHAIN TO MEASURE A STRAIGHT HORIZONTAL LINE

The ends of the line to be measured are marked by vertical ranging poles. Two people, called the

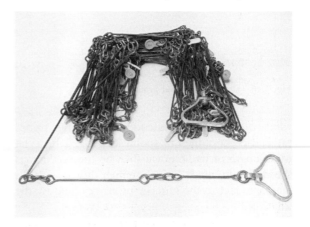

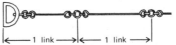

Fig. 2.14 Surveyor's chain.

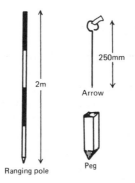

Fig. 2.15 Chaining: ancillary equipment.

leader and the follower, are required, one for each handle of the chain. The leader, who should have a ranging pole and a supply of arrows, sets off down the line with the chain until it is fully extended. The follower directs the leader on to the line by instructing him to move until the ranging pole is in line with those at the end points. With the handle of the chain firmly on the starting point of the line, the chain is shaken and pulled tight by the leader until it is just touching his ranging pole and is thus on line. An arrow is placed in the ground by the leader at the end of the chain.

The chain is then moved along the line until the follower reaches the arrow left by the leader. The follower puts the end of the chain against the arrow and the alignment procedure is repeated. When the follower leaves this station, the arrow is lifted. The

Fig. 2.16 Chaining on level ground.

whole operation is repeated until the leader passes the end-point of the line (point B in Fig. 2.16). The total length of the line will then be the reading on the chain at point B, plus a number of whole chain-lengths equal to the number of arrows held by the follower.

A similar procedure could be used with a tape or a steel band.

2.2.2.2. USE OF THE CHAIN/TAPE IN NON-FLAT TERRAIN

For plotting on a map to be possible, the horizontal distance between points must be known. On horizontal ground, this is obtained by simply reading the chain or the tape. With non-flat terrain, different procedures must be adopted.

If the terrain is plane but sloping, i.e. even slopes, the slope distance can be measured by normal procedures and converted to the horizontal distance using the angle of slope or the height difference between the points (Fig. 2.17). The angle of slope can be measured using a clinometer (sect. 2.3.1.1), or the height difference can be determined by a level or hand level (sect. 2.3.4). If the angle of slope is less than 3–4°, this conversion is not usually necessary and the slope

distance can be used as the horizontal distance without seriously affecting the accuracy of the method.

When the terrain is rough, there will not be a sloping line along which chaining can proceed and therefore step or drop chaining must be employed (Fig. 2.18).

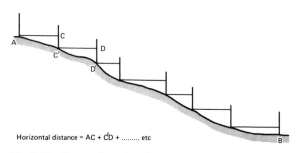

Horizontal distance = AC + C′D + ... etc

Fig. 2.18 Step or drop chaining.

In this case, the alignment is as before. The chain or tape is held horizontally in position AC and this horizontal distance is read. The position C is transferred to the ground at C′, using a ranging pole, a plumb bob or a drop arrow (a metal arrow with a lead weight attached to it, which can be held with the chain or tape at C and dropped to stick into the ground vertically below C to mark C′). The distance C′D is then measured, and so on. The lengths AC, C′D, etc. will usually be short in order to minimise the sag in the chain. The actual length used, which for convenience is usually a whole number of main divisions, will also depend on the steepness of the ground. It is useful to have a third person standing well clear of the line to check whether the chain is being held horizontally at the moment the reading is taken. This method is only successful if carried out with great care and should be avoided if at all possible as it is difficult to produce good results.

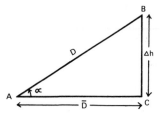

D = measured slope distance
$\bar{D}$ = required horizontal distance
Δh = difference in height between A and B
α = angle of slope

$\bar{D}$ = D cos α   if angle is measured

$\bar{D}$ = $\sqrt{D^2 - \Delta h^2}$ if height difference is measured

Fig. 2.17 Slope correction.

2.2.2.3 FIELD PROCEDURE FOR A CHAIN SURVEY

The field procedure for a chain survey can be broken down into four distinct phases.

1. Reconnaissance and choice of control figure.
2. Choice and marking of control stations.
3. Measurement of main lines.
4. Measurement of detail.

Reconnaissance

Detail measurement in a chain survey depends on the measurement of offsets from a main line. In this first phase, it must be established whether one line will be sufficient or if a number of main lines will be necessary. If the area to be mapped is small and somewhat linear in extent, all that need be done is to choose the main line through the area that passes closest to the detail to be mapped and is free from obstacles.

If the area to be mapped is not restricted to a narrow band of territory, more than one line will be required and a network must be established. A good network for a chain survey must satisfy the following conditions.

1. It must be a strong geometrical figure with at least one check line, e.g. a braced quadrilateral (see Figs. 1.1 and 2.24).
2. All detail to be surveyed must lie close to one or other of the main lines.

Choice and marking of the control stations

Once the approximate positions of the main lines have been established, the control stations marking the ends or intersections of lines can be chosen. Care has to be taken in their choice if intervisibility is to be maintained.

As these stations are required throughout the survey and may be needed to allow an extension or revision of the survey, they should be well marked. An artificial mark such as a peg, or a natural feature such as a fence post, drain or kerb stone may be used, but in all cases a sketch should be made showing the location of the point with respect to fixed surrounding detail (Fig. 2.19).

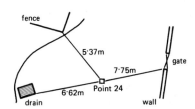

Fig. 2.19 Point description diagram.

Measurement of main lines

It is normal to measure main lines twice in opposite directions to check the field work. In many surveys it is not possible to choose main lines that do not

contain some obstacles to measurement or intervisibility. Some common problems and their solutions are given in Fig. 2.20.

a)

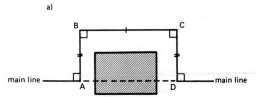

If the angles at A, B, C and D are all 90°
and if AB = CD, then the distance AD (the break
in the main line) will be equal to BC

b)

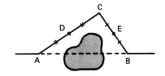

The distance DE, between the mid-points of
AC and BC, will be equal to half the distance AB

c)

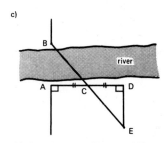

With the angles at A and D equal to 90° and
C the mid-point of AD, then if E is in line
with C and B, DE will be equal to the distance AB

d)

Section

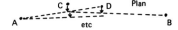

When the line of sight between two end points is
obstructed by a hill, two points on the hill on line
can be found in the following way:

Place two ranging poles as near as possible on line
such that both can be seen from A and B. Using A
and D, bring C on line. Using B and C bring D on
line.

Repeat until no further movement is necessary,
when A, C, D and B will be collinear

Fig. 2.20 Some common problems in chaining and their solutions.

a)

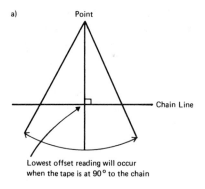

Lowest offset reading will occur when the tape is at 90° to the chain

b)

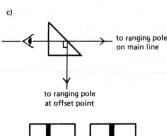

When the tape is pulled tight, a right-angled triangle will be formed

c)

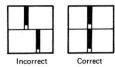

to ranging pole on main line

to ranging pole at offset point

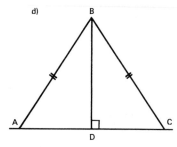

Incorrect   Correct

When the two poles are in line in the sighting device, the optical square is over the point on the main line from which the offset should be measured

d)

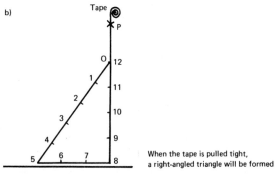

If AB = BC, then D at the mid-point of AC is the point from which the offset to B should be measured

Fig. 2.21   Methods of laying out a right angle.

Measurement of detail

Topographic detail is measured by offsets. The chain is laid along the main line and offsets are measured by tape. It is possible to carry out offset measurements when the main lines are being measured, but it is wise to measure and check the main lines (the framework) before starting the detail work.

The right angle necessary for an offset measurement can be laid out in a number of ways, as shown in Fig. 2.21.

For short offsets (less than 3–4 m), the angles can be set off by eye. For offsets up to 6 m, the right angle can be checked by swinging the tape around the end mark and reading the chain value at the shortest offset value (Fig. 2.21a). A 3, 4, 5 triangle can be constructed with the tape and moved along the chain until it is in line with the point of detail (Fig. 2.21b). In this example, 8 m would be subtracted from the offset as read on the tape to give the true value. An optical square could also be used. Standing on the chain line, and moving the prism along the line until the point to be offset is seen at the same time as the ranging pole at the far end of the line, will ensure a 90° offset (Fig. 2.21c). Two equal lengths of tape can be used to join the point to the main chain line such that the angles at the line are less than 60°. The offset is then measured from the point on the chain midway between the positions on the chain of the two tape ends (Fig. 2.21d).

If there is no clear line at 90° from the chain to the point to be surveyed (because of an obstruction such as a tree), offset measurement can be replaced by trilateration. The distance from the point is measured from two points on the chain (normally at distances which have convenient round number values), and the point can be plotted by scaling off these two distances (Fig. 2.22).

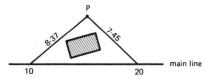

Fig. 2.22   Trilateration to replace an offset measurement.

Zero offsets (where a main chain line crosses a line of interest or point of detail) should also be recorded.

2.2.2.4 BOOKING THE MEASUREMENTS

In all survey work, it is important to record the field observations in a clear and systematic manner. It is a useful concept to visualise the booking sheet as a miniaturisation of the field. If readings are booked in such a way that a draughtsman who has not been in the field, and who cannot consult the surveyor for advice, can carry out the plotting, then a satisfactory method of booking is being employed.

A figure should be drawn, approximately to scale, showing the positions of the main lines and the control points should be given names, numbers or letters for identification. The main line measurements should be noted with the mean accepted value clearly indicated.

The booking of the offset measurements is carried out in a special format (Fig. 2.23).

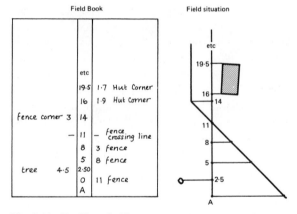

Fig. 2.23   Booking of offset measurements.

The distances along the main lines are noted within the central columns, with the offset distances at each side. Normally the booking is started at the foot of each page so that offsets to the left and right of the chain appear to the left and right of the page as the booker moves along the line. There is room for a brief description of the points measured and small sketches can also be included. The practice of making small field sketches is strongly recommended for this and most other survey operations.

2.2.2.5 PLOTTING THE SURVEY

The first task is to plot on a plotting sheet the positions of the control points at the required scale. If coordinates are available for two of the control points used in the control figure, a coordinate grid should be drawn up on the sheet and the points plotted (see Fig. 2.4).

If no coordinates are available, the survey will have to be plotted in a local system based on one of the control lines (Fig. 2.24).

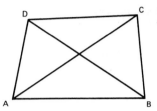

A good control figure for a roughly rectangular area is a braced quadrilateral

Fig. 2.24   Control figure.

Let this line be AB, which can be plotted on the map sheet in one of two ways. Assuming that AB is 125.37 m in the terrain and the map scale is to be 1:500, the distance between A and B on the map sheet is obtained by dividing the field measurement by the map scale number, i.e. by 500.

Terrain distance = 125.37 m.
Map distance   = 125.37/500
          = 0.2507 m
          = 250.7 mm.

Alternatively, a rule graduated with a 1:500 scale may be used to plot directly the correct distance.

With the line AB acting as a base line, the positions of points C and D can be plotted by trilateration (Fig. 2.25).

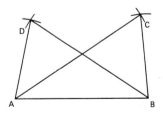

Fig. 2.25   Plotting the control figure.

The distance AD is set on a beam compass (Fig. 2.26) and an arc is drawn in the region of the position of point D. Using the distance DB, a

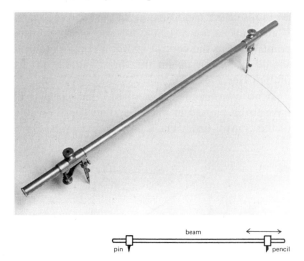

Fig. 2.26   Beam compass.

similar arc is drawn from B to meet the one drawn from A. The intersection of these two arcs gives the position of the point D. A similar procedure is followed to plot the position of point C.

At this stage there is the possibility of an independent check for, although the positions of the four points have been established, one field measurement has not been used. This distance (CD) is converted to map scale and set on the beam compass and compared with the plotted distance between the two points C and D. If these two values agree, all is well, for an independent check on the work so far has been provided; if the two values differ by more than an acceptable margin, an error has occurred and it must be located. However, the surveyor should be optimistic and should check the plotting of the points and the reduction of the field book first before deciding that observations need to be taken again in the field.

If the plotting is to be carried out at a large scale, it may be better to compute the coordinates of the four control points and plot these points on a gridded sheet. An example of a trilateration computation is given in section 2.6.2.6.

Once the plotting of the control points has been completed satisfactorily, the plotting of the detail can begin. A graduated scale is laid along one of the control lines, with the zero mark against the control point. A small offset scale is then moved along the main scale until the point from which the first offset measurement was taken is reached. The point of detail can then be plotted, using the

(a)

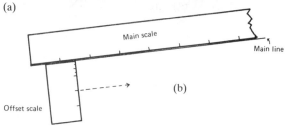

(b)

Fig. 2.27   (a) Use of offsets on an archaeological site.   (b) Use of the offset scale.

graduations on the offset scale. This operation is repeated along the length of the main line (Fig. 2.27b). It is convenient when using small offset scales to plot all offsets on one side of the main scale first rather than changing from side to side, as is normal during field work.

During plotting, the draughtsman should be on the lookout for possible errors. If an otherwise straight line or smooth curve plots out with a displacement or a kink, an error should be suspected and checks should be carried out.

2.2.2.6 APPLICATIONS OF TRILATERATION AND OFFSET BY CHAIN AND/OR TAPE

The advantages of this method are that the equipment is relatively cheap, rugged and easily used. Also, the field procedures are simple. In reasonably flat, open ground, results come very swiftly, but in broken or wooded terrain chaining and taping become difficult and accuracy is lost through frequent use of step chaining. With long distances, especially when little detail has to be picked up, this method is much slower than others. With double chaining of the main lines and with reasonable care in the measurements of the offsets, an accuracy of one part in 500 can be obtained with this method.

The following are suitable applications.

1. Where the survey area is small in extent but requires mapping at a fairly large scale, e.g.: archaeological sites (Fig. 2.27a); abandoned settlements; soil, vegetation or other boundaries within a transect; or buildings.
2. Where a grid has to be established over a piece of ground for the purpose of sampling or monitoring changes, e.g.: micro studies in vegetation, soils and ground water studies; establishing a grid for selective excavations on an archaeological site; or as a base for contour mapping (see sect. 2.4.3).
3. Where transects have to be run, and where the significant features are all close to one line. This

work can be carried out in combination with levelling.

2.2.3 TRIANGULATION BY PLANE TABLE

When the area to be mapped is larger than can be conveniently tackled by the chaining method, the graphical triangulation system of plane table surveying can be considered. Angles, not distances, are employed in this method, which can be carried out more easily where long distances have to be covered, especially when only simple equipment is available.

2.2.3.1 EQUIPMENT

A plane table is essentially a drawing board that can be mounted horizontally on a tripod by means of a central pivot about which it can be rotated and then fixed in any desired position using a clamping screw. With a sighting device called an alidade, the surveyor carries out a graphical triangulation in the field, thus fixing points of interest without actually measuring the angles. Other useful items of equipment are: (1) a box or trough compass, which can be used to orientate the board with respect to magnetic north; (2) an Indian clinometer, which is used to determine the height difference between points; and (3) chains and tapes if radiation and traverse techniques are also to be employed. Figure

Fig. 2.28 Plane table and associated equipment.

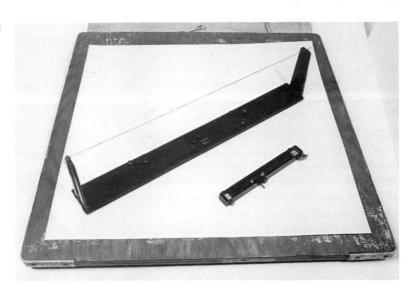

2.28 shows a plane table with an alidade and box compass.

Except in very small surveys or in exploratory surveys where accuracy may have to be sacrificed for speed, graphical triangulation on the plane table should be dependent on a number of control points whose coordinates or relative positions are known. In other words, a control network must be established before the detailed mapping begins. Again, this is an application of the principle of working from the whole to the part. A good network of points throughout the area ensures that the errors inherent in a graphical triangulation will not accumulate to such an extent that surveyed detail will be displaced from its true position by an amount that is plottable at the scale of the survey.

A grid is therefore drawn on the base sheet and all points of known coordinates are plotted on the sheet at the scale of the survey. These points should be plotted in ink in order to avoid smudging during field work.

As the plane table will remain for some considerable time in the field, sometimes in unfavourable weather conditions, it is essential that the material used for the plotting sheet should be stable, even under varying conditions of temperature and humidity. Originally, linen-based paper sheets were used, but now aluminium-backed sheets such as 'Correctostat' and plastic drawing sheets are usually employed. A hard pencil (2H/3H) and a soft eraser are required (see the section on pencil drawing).

2.2.3.2 FIELD PROCEDURE

The procedure for plane table triangulation can be discussed under three headings: setting up and orienting the board; intersection; and resection.

Setting up the board at a control point (known point)

In small-scale work it is not necessary to set up the plane table very accurately over the control point on the ground, for the error in centring will be smaller than can be plotted at the scale of the map. At large scales, however, it is very important that centring be carried out accurately.

The control point will be marked on the ground, usually by a peg or a bolt, and the point on the plot

sheet representing that control point must be vertically over that mark. It will usually be possible to orient the board, within say 10–20°, by comparing the layout of points on the board with the corresponding points in the field. This should be done and the board and tripod moved until the plotted point is as near as can be judged over the mark. In order to check this position for large-scale work, or to determine the further movement necessary, a plumbing bar can be used. This bar is a hairpin-shaped device which has arms of equal length, one ending in a pointer, the other in a hook to which a plumb bob is attached (Fig. 2.29).

Fig. 2.29 Use of a plumbing bar.

The plane table is in the correct position when the pointer is on the control point and the plumb bob is over the mark on the ground. The levelling of the table can then be checked by eye, from two directions separated by 90°, or by a level bubble. The table is now ready for orientation.

Orientation by means of control points

With the table set up at point A (Fig. 2.30), a second control point visible from (but distant from) A is chosen (say B) and the alidade is laid along the line AB on the table. The vertical axis clamp is then released and the table is rotated until point B in the field is seen through the sight of the

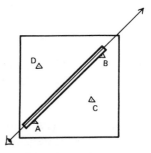

Fig. 2.30 Orientation of the board by means of an alidade.

alidade. The table is again clamped and, as line AB in the terrain is now coincident with line AB as plotted on the map sheet, the table is correctly oriented.

If the preliminary orientation was bad, or if point A is very far from the centre of the board, the station point may no longer be over the ground mark and this must be checked. An iterative procedure may have to be employed in large scale work in order to obtain correct orientation and centring.

With the orientation complete, it is useful to add a magnetic north indication to the board. This is done by placing the box compass on the board and rotating it until the needle is on the zero mark. A line is then drawn along the edge of the box and this can be used to orientate the board using the method described below.

Approximate orientation using a box or trough compass

The compass can only be used if a magnetic north indication has already been plotted on the sheet, either during the preparation of the sheet or using the method described above.

The compass is laid along the magnetic north indication and the board is unclamped and rotated until the needle covers the north indication. The board is then clamped and orientation has been achieved. This is a much less reliable method than the one previously described and should only be used to obtain a preliminary, approximate orientation of the board.

Using the principle of triangulation, the two methods of fixing additional points are intersection and resection.

Intersection

Intersection is the method of fixing in which angular observations are taken at known points to fix the position of unknown points. The field procedure is as follows.

1. The plane table is set up and levelled at the first control point (A in Fig. 2.31), and the board is oriented using, for example, line AB.
2. The alidade is pivoted around A until point P can be seen in the alidade sight. The position of P must lie on the line defined by the alidade. At this stage, only a short part of the line is drawn

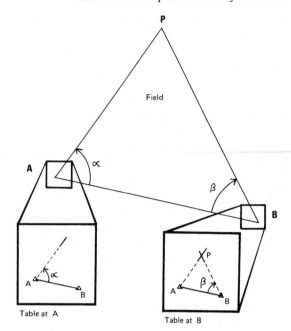

Fig. 2.31 Plane-table intersection.

at the edge of the board. This operation records graphically the angle from the base line AB to point P.

3. The plane table is then moved to control point B, and is then centred, levelled and oriented.
4. The alidade is pivoted around point B until P is sighted, and again a small section of this line is marked at the edge of the board. This operation graphically records the angle from the base line BA to point P.
5. The position of P on the plot can now be determined by finding the intersection point of these two rays.

Obviously, all the sightings required from point A would be taken before moving to point B. In order to avoid confusion, all rays must be clearly labelled, to enable the matching of correct pairs of rays in the intersections. If many rays are to be taken from a point, the orientation of the board must be checked from time to time by sighting to the orienting point, and the last observation from a point should always be a check on the orienting ray.

If point P is to be used as a plane-table station later in the survey, it should be intersected from a third point to give an independent check on its position.

29

Resection

Whereas intersection is a method of determining the position of an unknown point by taking observations at known points, resection depends on observations taken at the unknown point. The main problem in resection is orientation. One solution is to take a forward sight from a known point before moving to the unknown point which has to be fixed. This is known as resection after setting the back ray. The procedure is as follows.

1. The plane table is set up, levelled and oriented in the usual manner at point A (Fig. 2.32).
2. A sight is taken to the unknown point P, as was done in intersection. This direction is recorded at the edge of the board. The position of P must lie somewhere on this line.
3. The plane table is set up at point P and is levelled. Although the position of P is not known, the direction from P to A is known and therefore, by aligning the alidade along PA and rotating the board until A is visible in the alidade sight, the board can be oriented.
4. As the plane table is oriented, the line joining B in the terrain with its plotted position on the board must also pass through P. The alidade is therefore pivoted around B on the board until B in the terrain is seen in the sight. This line will

also contain P and is marked at the edge of the board.
5. The plotted position of P will be found at the intersection of the back ray (from A) with the ray from B. A third ray should be used to confirm the position of B.

This method is really a combination of resection and intersection, orientation being achieved by the ray put into the unknown point from a control point.

Normal resection is carried out by taking observations at the unknown point only, i.e. no back rays are available to orient the board. The surveyor must be able to see at least three points whose positions are already plotted on the board. In this type of resection, the unknown point may fall within the triangle formed by the three control points (Fig. 2.33a) or it may not (Fig. 2.33b).

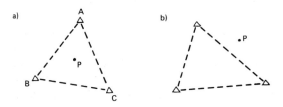

Fig. 2.33   Two types of resection.

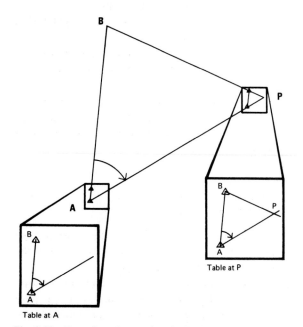

Table at P

Table at A

Fig. 2.32   Resection after setting the back ray.

Unknown point within control triangle

The field procedure is as follows.

1. The plane table is set up and levelled over the unknown point.
2. The board is oriented as well as possible, using either a trough compass or by comparing the position of the control points plotted on the board with the corresponding points in the field.
3. The alidade is pivoted around one of the control points on the board until that point can be seen through the sight. A ray is drawn back from the control point to the region of P on the board. This is repeated for the other two control points.
4. If the board has been oriented exactly, the three rays should meet in a point. But with any orientation error (and this is likely), the rays will form a triangle, known as the triangle of error, within which lies the true position of P.
5. A provisional position of P is chosen within the triangle such that the perpendicular distances

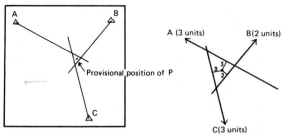

Fig. 2.34 Resection, Case 1.

from the rays to P are proportional to the lengths of the rays.

6. The board is re-oriented using this position of P and the most distant control point, and the resection is repeated until the three rays meet in a point (Fig. 2.34).

Unknown point outside control triangle

The rays are drawn in as described in steps 1–4 above, and once again a triangle of error will be found if the orientation of the board was not correct. This time the true position of P will lie outside the triangle. As any alteration of the orientation of the board will cause all the rays to swing in the same direction, the true position of P must lie to the left or right of all the rays. The area in which P lies is found by considering the perpendiculars to the rays (Fig. 2.35).

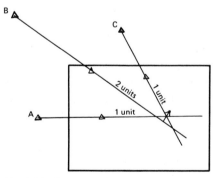

Fig. 2.35 Resection, Case 2.

The position of P is estimated, the board is re-oriented and the resection repeated until the three rays meet in one point.

When resection is being used to fix the position of a point lying outside the triangle formed by control points, there is a chance that these control points and the unknown point may all lie on the circumference of the same circle. In such a case, no triangle of error will be formed, even if the orientation of the board was in error. If the first plot of the position of the unknown point appears to lie close to the circle through the control points (known as the danger circle), it is necessary to replace one of the control points with another well clear of this circle, thus avoiding the problem.

For both types of resection, the final position of the unknown point should be checked by taking rays to it from all the visible control points.

Bessel's method of resection

If the plane-table board can be accurately aligned and clamped, the resection method of Bessel is recommended. The procedure is as follows.

1. With three known points (A, B and C) plotted on the board (a, b and c), the plane table is set up, levelled and approximately oriented at the unknown point P.
2. The alidade is set along line ab on the board and the board rotated until the alidade points to B, when the table is clamped.
3. The alidade is pivoted around a to sight C, and the ray a to C is drawn.
4. The alidade is laid along the line ba and the table is rotated until the alidade points to A in the field. The table is clamped.
5. The alidade is pivoted around b to sight C and the ray b to C is drawn.
6. The alidade is laid from the intersection of lines aC and bC to point c and the table is rotated until point C is sighted.
7. As the table is now oriented, the resection can be completed by drawing rays back from a and b as described in the method of resection after setting the back ray.

Mechanical resection

Resection can be accomplished with the aid of a piece of tracing paper in the following manner.

1. The plane table is set up over the point and the board is levelled. The table need not be oriented, even approximately.
2. A dot is made in the centre of the tracing paper to represent the position of the unknown point and the tracing paper is fixed to the board with drafting tape.

3. The alidade is pivoted around the dot and rays are drawn to the three control points in the field.
4. The tracing paper is moved over the board until the three rays pass through the three control points plotted on the board. The dot then marks the position of the unknown point on the board (point p).
5. The dot is pricked through on to the plot sheet, the alidade is laid along the line from p to the most distant point and the table is oriented.
6. The resection can be checked by bringing in rays from other control points.

This method is commonly used in hydrographic surveying, where the position of the survey boat has to be determined by resection from angles measured by sextant. A mechanical device known as a station pointer is used to facilitate the plotting (sect. 2.5.3).

2.2.3.3 APPLICATION OF PLANE-TABLE TRIANGULATION

At large scales, greater than 1:2500, plane-table triangulation can be used to break down an existing control network to give a more dense coverage of lower-order control points, but at these scales other methods are more suitable for surveying detail. At medium to small scales, especially when an existing map is available, plane-table triangulation is a convenient method of plotting the position of prominent features of the landscape. Since the map builds up in the field under the eye of the surveyor, observations are not easily forgotten and verification can take place as the work progresses. With no computation or other office work to follow, the surveyor leaves the field with a map ready for use, or ready for the cartographic procedures. With experience, it is possible to sketch in rather than survey all the necessary detail and this allows the work to progress quickly. In wooded areas, line-of-sight problems may arise. One other limitation of the technique is that as the map is being produced in the field reasonable weather is essential.

Plane tabling is most suitable for adding point information to existing maps, either for revision or for the addition to regular maps of a distribution that is specially significant to the field scientist, e.g. vegetation boundaries, sample sites, or former settlements. When a map at a suitable scale is available, the detail on the map itself can be used as a control, the map being placed on the board and used as the plotting sheet.

Neatness is of great importance in this method, as many rays will have to be drawn on the plane-table sheet in the course of survey, and without due care the plot can become confused and smudged. Readers should consult the section on pencil drawing for hints on how to maintain a neat and legible plot (Ch. 5).

2.2.4 RADIATION BY PLANE TABLE

Although triangulation methods using the plane table are well suited for the establishment of control networks or for fixing a few points of detail over long distances, they are not so suitable for the determination of a large number of points close to the occupied station. With many rays, care has to be taken to match the correct pair of rays for intersection, and the points have to be marked in the field by ranging poles or arrows.

For large-scale surveys, close detail is best established by the method of radiation. This is based on the measurement of angles and distances, and the field procedure is as follows.

1. The plane table is set up at a control point and is correctly oriented (see sect. 2.2.3.2).
2. The alidade is pivoted around this point on the board to sight the first point of detail. The distance to the point is measured by chain or tape and is scaled off along the alidade to find the plotted position of the point.
3. This is repeated for all other points of detail.

After every tenth observation or so, the orientation of the board should be checked, lest the tripod has been moved slightly. Sometimes the weight of the operator leaning on the board is sufficient to disturb the orientation, especially if the clamp has not been tightened properly. The operator should only rest his finger tips on the board during the drawing of rays.

2.2.5 SIMPLE METHODS OF TRAVERSING

Traversing, which uses both angle and distance

measurement, is suitable for control surveys where measurement along a line is required, e.g. along a road or track, around the perimeter of a forest or along the banks of a stream. The detailed surveying can then be carried out by radiation, intersection or offsets from the traverse stations and legs.

Traverses which start and finish at points of known position are called closed traverses, while those which do not close on a known point are called open traverses. An open traverse is a very weak survey system, there being no check on the work, and may only be used as a last resort for picking up minor detail and should never be used for control surveys.

Closed traverses may return to their starting point (a loop traverse) or may finish on some other known point (a connecting traverse). In each case, a check on the work is available, as any accumulated error will be detected at the closing station. Only closed traverses will be considered in this section.

2.2.5.1 TRAVERSING WITH A COMPASS AND CHAIN

In this method of traversing, the distances are measured with a surveyor's chain, and the angles or bearings with a surveyor's compass or a prismatic compass. In a more elementary method, the distances could be paced.

The surveyor's compass is the larger of the two types (up to 200 mm in diameter). It is usually placed on top of a tripod and gives readings to 20 minutes of arc. The prismatic compass is a hand-held instrument, and readings are taken through a small 45° prism, which allows the view along the survey line and the compass dial to be seen simultaneously. Each type is manufactured in many forms, two of which are shown in Fig. 2.36.

Field procedure

1. All traverse points are marked by ranging poles.
2. The compass is used at the first point to determine the magnetic bearing to the second point. While this is being done, no metal object such as a steel band or chain should be allowed close to the instrument.
3. The distance to the second point is then measured in such a way that the horizontal

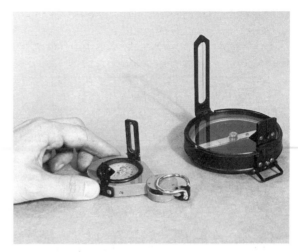

Fig. 2.36 Surveyor's compass and prismatic compass.

distance can be found for plotting. If necessary, step chaining should be employed or slope angles measured.

4. The compass is then moved to the second station. As a check against gross error, the back bearing to the first station can be read. This should differ by 180° from the forward bearing. Any small discrepancy can be halved and applied to the forward bearing. The bearing and distance to the new point are then measured.
5. This procedure is repeated for each leg along the length of the traverse until the traverse closes on to a known point.

It should be noted that it would be possible to use the compass at every second station only, the forward bearings from the unoccupied stations being obtained by taking 180° from the backward bearings at the occupied stations, but there would be no check on gross error. Further, since a compass is liable to several forms of error, it is normally advisable to read any compass bearing more than once.

Plotting the traverse (graphical method)
Assuming that the traverse runs between two points A and B, through traverse points 1, 2 and 3, a grid is drawn up and points A and B are plotted at the required scale (Fig. 2.37).

A north-to-south gridline is drawn through point A; this indicates the direction of north at point A. On the assumption that magnetic north is 7° west of grid north, a circular protractor is centred on A

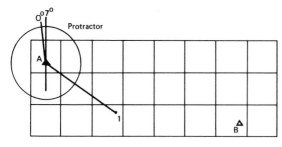

Setting up a protractor at A, with the value of the
magnetic declination on a N–S line, the position of 1 can be determined
by scaling off the distance A 1 on the appropriate bearing from A.

Fig. 2.37 Graphical plot of a traverse.

and is rotated until the 7° mark coincides with the
north grid line. The protractor can then be used to
plot the magnetic bearings from A. A mark is made
on the plotting sheet against the value of the
magnetic bearing to traverse point 1. The
protractor is removed and a scale is lined up with
this point and A, with the zero point of the scale
at A. The distance from A to point 1 is scaled off,
thus determining the position of point 1.

A north-to-south line is drawn through the
plotted position of point 1 and the procedure is
repeated to plot the position of point 2. This is
repeated along the length of the traverse until the
final point (the closing point) is plotted. If all has
gone well, the plotted position of point B should
coincide with the control point position. Usually,
however, a small discrepancy will occur and this
must be removed by adjusting the position of all
the points in the traverse (Fig. 2.38). With the final
plotted position at B′ instead of B, the distance
BB′ is called the misclosure. The adjusted positions
of the points are found in the following way.

1. B′ and B are joined and the distance BB′ is
   measured.
2. Lines parallel to BB′ are drawn through all the

other traverse points. The adjusted position of
all the other traverse points will lie on these
lines.
3. The movement from the original position of each
   point to its new position is found graphically
   (Fig. 2.38).

A compass and chain traverse can be used to fix
control for an offset survey in an area where woodland, buildings or other obstacles prevent trilateration. It must be remembered, however, that
traversing is not a method in which a strong
geometrical figure is possible, and the chance of
errors occurring and remaining undetected is high.
It is for this reason that traverses must always run
between points of known position.

2.2.5.2 TRAVERSING WITH A PLANE TABLE

Just as traversing can help in areas which are
difficult for trilateration, the method can also be
applied in areas causing difficulty for triangulation.
In areas where all-round vision is restricted, there
are many problems connected with the formation
of triangles with good intersection angles. When a
control network has to be extended along a linear
feature such as a road, or when visibility is
restricted on either side, triangulation is often
impossible. The following example illustrates the
way in which traversing can supplement control
established by plane table surveying.

Figure 2.39 shows an area where plane-table
triangulation has been used to establish a network
of points which must now be extended into the
clearing in the forest. This can be achieved as
follows.

1. The plane table is set up at point A and oriented
   using the line AB.

**Fig. 2.38 Graphical adjustment of a
compass traverse.**

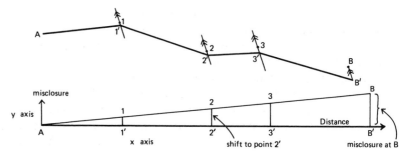

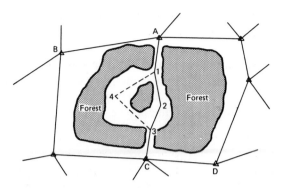

Fig. 2.39   Plane-table traversing.

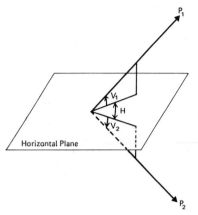

Fig. 2.40   Angular measurements with a theodolite.

2. The alidade is pivoted around A on the board until traverse point 1 can be seen in the sight. A line from A in the direction of point 1 is marked at the edge of the board.

3. Using a chain or tape, the distance from A to traverse point 1 is measured. The distance is scaled off on the line already marked and thus the position of traverse point 1 on the board is established.

4. The board is then moved to point 1, oriented on line 1 to A and a ray and distance measured to point 2.

5. This procedure is repeated until the traverse again meets a control point (e.g. point C). Any misclosure is adjusted in the way described for the compass traverse. Note that point 4 could be intersected from the final positions of points 1 and 3.

2.2.6 THE THEODOLITE AND ITS APPLICATION TO PLANIMETRIC SURVEYING

The theodolite is the most versatile, best known and most extensively used of all survey instruments. All the survey manufacturers produce a range of theodolites, differing in size, accuracy and cost. But for all their differences, all theodolites are designed to perform two simple operations, namely to measure angles in horizontal and vertical planes.

If directions P_1 and P_2 in Fig. 2.40 are the directions to points P_1 and P_2, the vertical angles (measured above or below the horizontal plane) to the two points are V_1 and V_2 and the horizontal angle between them is H.

2.2.6.1 THE THEODOLITE

A theodolite, at its most basic, is an instrument containing two 360° protractors, one mounted horizontally, the other vertically, against which two index marks move, allowing two readings to be taken for each pointing of the viewing telescope.

The main components of a typical theodolite are shown in Fig. 2.41.

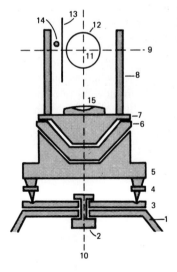

Fig. 2.41   Main components of a theodolite.

The trivet plate(3) is attached to the top of the tripod(1) by a clamping screw(2). The tribrach(5) rests on the trivet on three footscrews(4), which can be used to level the tribrach and the whole upper part of the instrument by reference to the horizontal or plate bubble(15). Above the tribrach are the lower plate(6) and upper plate(7), which are force-centred on the vertical axis(10) by their shape. Two standards(8) support the horizontal or trunnion axis(9) on which is fixed the telescope(12). The optical axis of the telescope(11) coincides with the intersection of the vertical and trunnion axes. Also on the trunnion axis is the vertical circle(13) with an index mark and bubble(14). Clamps and slow-motion screws are available for the upper and lower plates and for the movement of the telescope around the trunnion axis, and an adjusting screw is available for the vertical circle index bubble.

In older instruments, an external focusing telescope was used, but newer instruments have internal focusing. The image is brought to the plane of the diaphragm where a crosshair pattern is engraved, thus allowing accurate pointing. Some typical diaphragm patterns are shown in Fig. 2.42.

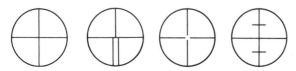

Fig. 2.42   Typical diaphragm patterns.

The circles on modern theodolites are divided either according to the sexagesimal system (one revolution = 360°) or the centesimal system (one revolution = 400 grades) and, whereas metal was formerly used, glass circles are now more common. There are five possible reading systems, examples of which are given in Fig. 2.43, and full details are always given in the manufacturer's instruction booklet and will not be discussed further here.

2.2.6.2 SETTING UP A THEODOLITE AT A POINT

1. Using a plumb bob, the tripod is centred over the point. The top of the tripod should be as level as possible and the tripod legs secured firmly in the ground.
2. The instrument is placed on top of the tripod and the trivet plate clamped to the tripod by means of the clamping screw.
3. The main plate bubble is brought parallel to any two footscrews and the bubble is moved to the centre of its run by turning the two screws

together but in opposite directions (Fig. 2.44a). (Note that the bubble will move in the direction of the left thumb – a genuine rule of thumb!) The instrument is turned through 90° and, using the third footscrew, the bubble is centred (Fig. 2.44b). The instrument is turned back through 90° until the bubble is again parallel to the first two footscrews (Fig. 2.44c) and the levelling procedure is repeated until the bubble is central in both positions.
4. When the instrument is level, the plumb bob or optical plummet is used to ensure that the vertical axis of the theodolite is directly over the peg marking the point on the ground. If not, the clamp is loosened and the instrument moved over the tripod plate until central. Re-levelling may be necessary and steps (3) and (4) should be repeated until the instrument is level and over the point.
5. The telescope is then pointed to the sky or a piece of white paper held in front of the objective lens. The eyepiece is focused until a sharp image of the crosshairs can be seen.
6. Looking through the reading telescope, the positions of the mirrors are adjusted so that the reading scales are well illuminated.

The instrument is then ready for use.

When the telescope is pointed towards the object to be sighted, the image of the object must be brought into focus using the focus screw. As objects to be sighted will be at varying distances from the instrument, the focus will need adjusting at every pointing. The eyepiece focus should not require adjustment unless a different observer comes to use the instrument.

It takes practice before good focus is achieved. If in doubt, the observer should move his eye from side to side across the eyepiece lens. If the scene in the telescope shifts with respect to the crosshairs, correct focus has not been achieved and further adjustments are necessary.

2.2.6.3 MEASUREMENT OF HORIZONTAL ANGLES

The use of a theodolite to determine the horizontal angles between points will now be discussed. Assume that the theodolite is at station PARK and observations have to be taken to stations SPIRE, POLE and TREE (Fig. 2.45).

SPECIMENS OF THEODOLITE SCALES

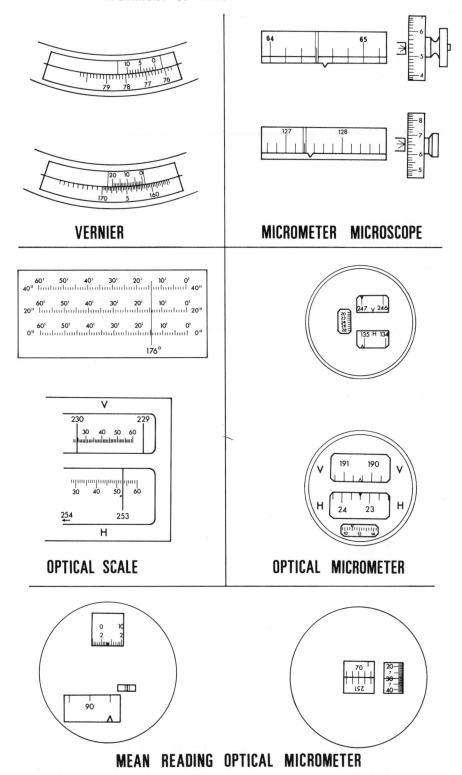

Fig. 2.43 Specimens of theodolite scales.

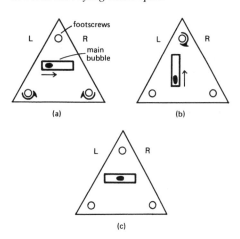

Fig. 2.44 **Levelling an instrument by means of three footscrews.**

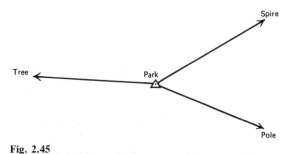

Fig. 2.45

1. One point to be sighted should be taken as the reference object (RO). This should be a fairly distant but well-defined point. If desired, an approximate value for its bearing (obtained from a compass) or a zero value can be set on the reading scale for the RO. The way in which this is achieved varies from instrument to instrument but details will be given in the instruction manual.
2. All points to be sighted are identified and listed in the order in which they appear in a clockwise direction, starting with the reference object.
3. The upper part of the theodolite is turned around a few times to take up backlash and the telescope is pointed to the RO on face left (FL), i.e. with the vertical circle on the left of the telescope. The horizontal and vertical movements are then clamped.
4. The vertical crosshair is brought on to the object with the slow-motion screw, making sure that the last motion is against the spring. The horizontal reading is then noted.

5. This is repeated for all stations, working around in a clockwise direction.
6. The telescope is turned around the trunnion axis and pointed to the last observed station on face right (FR), i.e. with the vertical circle to the right of the telescope. The FR horizontal scale readings are taken.
7. This is repeated for all stations back to the RO, working in an anticlockwise direction.

NB. *All horizontal readings on FL are taken before starting the horizontal readings on FR, which are taken in the opposite order.*

2.2.6.4 BOOKING OF HORIZONTAL READINGS AND COMPUTATION OF ANGLES

The booking of the readings and the subsequent computation of the angles should be carried out on a table. There are many methods of laying out such a table; only one is illustrated here (Table 2.1).

In Table 2.1, it can be seen that the readings are entered in column 4 in pairs. If the theodolite is in perfect adjustment and no errors in pointing or identification occur, the difference between corresponding FL and FR observations should be 180°. Owing to instrumental and observational errors, this difference will normally vary slightly from 180°. It can be proved that many of the instrumental errors will be eliminated if the mean of FL and FR observations are used, and small accidental observational errors tend to be reduced by taking a mean. The degree value from the FL observation is retained but the seconds, and if need be the minutes, of the two are meaned and entered in column 5. By taking the difference between the mean readings to the points, the angles between the rays to the points can be determined (column 6). If the bearing of one line is known (say 0° for line PARK to SPIRE), the bearing of the other lines can be computed simply by adding the clockwise angles.

All theodolites are precision measuring instruments and should be treated with care. However, it is inevitable that these instruments will be subjected to some rough handling in the field and the manufacturers take this into account in the designs. Even with care, a theodolite will go out of adjustment and it will therefore need to be checked from time to time.

Table 2.1 Booking sheet for horizontal angles

1 Station	*2* To	*3* Face	*4* Reading	*5* Mean	*6* Clockwise angle	*7* Direction
Park	Spire	FL	02 10 20	02 10 25		00 00 00
		FR	182 10 30			
					63 14 10	
	Pole	FL	65 24 40	65 24 35		63 14 10
		FR	245 24 30			
					155 15 40	
	Tree	FL	220 40 10	220 40 15		218 29 50
		FR	40 40 20			
					141 30 10	
					360 00 00	

Check and adjustment procedures are outside the scope of this text, but details can be found in most standard survey texts. A field scientist owning a theodolite might persuade a surveyor or a manufacturer's agent to check the instrument periodically, if he wishes to avoid this task himself.

Most of the consequences of bad adjustment, however, are eliminated if the means of observations on FL and FR are used, and this procedure should always be adopted in control surveys. For detail survey, where the highest accuracy is not normally required, observations are usually taken on FL only and it is assumed that the instrument is either in good adjustment or that the consequences of poor adjustment are negligible at the scale of the plot.

2.2.6.5 APPLICATIONS

A theodolite can be used to measure horizontal angles for a triangulation scheme, for intersection, for resection, for radiation or for a traverse; in such cases the theodolite replaces the plane table and alidade in the methods already discussed. If graphical methods are to be used for plotting (usually a protractor and scale), angles need not be read to a high accuracy and readings on one face will suffice. Radiation using a theodolite and tape can be a very efficient method of producing a detailed survey at a large scale, for example, on an archaeological site.

A theodolite can also be used to set out lines prior to a survey. It may be convenient to cover an area with a regular grid of points within which an offset survey will be carried out. A theodolite can be used to set out lines at right angles very readily.

If a long line has to be chained, it is often easier to keep the chain on line using a theodolite set at one end than by using ranging poles. The same procedure could be used for setting out transect lines up hillsides or across beaches.

The full potential of the theodolite in planimetric mapping is not realised unless the positions of the unknown points are to be determined numerically, as usually happens in the control survey stage. For higher accuracy work, see section 2.6 where computational methods and the construction of frameworks are outlined.

2.3 HOW TO OBTAIN SLOPE AND HEIGHT VALUES

The problem of heighting can occur in several forms in field survey practice. The gradient of a

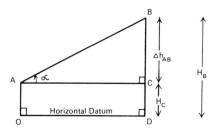

Fig. 2.46 Height relations.

slope, the relative height of one point to another, the absolute height of a point and the vertical dimension of an object are all variations on a 'triangular' situation.

If A and B in Fig. 2.46 are the two points under consideration, the gradient of line AB can be obtained from the angle of slope or the ratio of the distances BC and AC. The height difference between A and B is the vertical distance CB and the absolute height of A and B can be found from OA and DB, the vertical distances from the datum plane OD. Thus, if any two quantities in the triangle ABC can be determined, all other quantities can be found by simple trigonometry. The absolute height can only be introduced if the absolute height of A or B is known. This information can be obtained from bench marks (see sect. 2.3.4.3).

The possible methods of solving this triangle can be grouped into: (1) those based on slope angle measurement; and (2) those based on height difference measurement, where the height difference is measured directly or indirectly.

2.3.1 MEASUREMENT OF SLOPE ANGLES

In many field problems, the slope angle or gradient of a line is more meaningful than the height difference between its end points. This might be the case in geological or geomorphological studies, where the dip of strata or the gradient of a scree slope has to be measured.

In such situations, the survey problem is relatively simple. By using a clinometer, Abney level or theodolite, the vertical angle can be measured directly. Since the slope angles are often not required to any high degree of accuracy, the small

hand-held instruments can usually be employed, and after a little practice results correct to between 0.5° and 1° can be obtained.

2.3.1.1 CLINOMETER

The simple clinometer consists of a rough sighting device and a horizontal or vertical index, which is given either by a spirit bubble or by a plumb bob (Fig. 2.47).

Fig. 2.47 A simple clinometer.

To measure a slope angle, a pole is held at one end of the slope with a target at the height of the observer's eye, and the observer sights this target and takes the reading. This procedure ensures that the vertical angle measured is equal to the slope of the terrain. Over long distances a target is not necessary, but it essential for good results at short distances (Fig. 2.48).

2.3.1.2 ABNEY LEVEL

An Abney level gives more accurate results (10 to 15 minutes for arc) than a clinometer. This instrument (Fig. 2.49) is a combination of a hand level (sect. 2.3.4.1) and a simple clinometer.

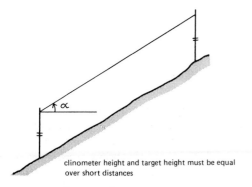

Fig. 2.48 Use of a clinometer.

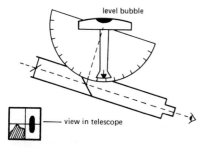

Fig. 2.49 Abney level.

On looking through the sighting tube, one observes two images – the point to which the angular measurement is to be taken and an image of the bubble reflected by the mirror set at an angle of 45° to the line of sight. With the crosshair on the target, the bubble should be brought to the centre of its run by means of the bubble screw. Attached at right angles to the bubble tube is an arm with a pointer and vernier scale, allowing a reading to be taken on a semi-circular scale.

2.3.1.3 THEODOLITE

When vertical angles are required to a higher degree of accuracy, the most suitable instrument is a theodolite. Its use to measure vertical angles will be discussed in detail in the following section. (Further details on the use of a theodolite can be found in section 2.2.6.)

2.3.2 TRIGONOMETRIC HEIGHTING

Trigonometric heighting is an indirect method of

obtaining height differences, since in the field it is the vertical angle, not the height difference, which is measured. From the heighting triangle already discussed (Fig. 2.46), the following relations can be deduced.

$$\Delta h_{AB} = AC \tan \alpha$$

or

$$\Delta h_{AB} = AB \sin \alpha$$

The vertical angle α can be measured by one of the instruments discussed previously, and the accuracy of the height differences obtained will depend on the accuracy of the vertical angle and on the distance measurement (either the horizontal distance AC or the slope distance AB). For rough height differences over short distances, the clinometer or Abney level will suffice, but for high accuracy over long distances a theodolite must be employed.

2.3.2.1 THEODOLITE MEASUREMENT OF VERTICAL ANGLE

Assuming that the height differences between PARK and points SPIRE, POLE and TREE in Fig. 2.45 are required and that the horizontal distances between PARK and these three points are known, the observational procedure can be similar, but not identical, to that for horizontal angles. In particular, three additional points must be considered.

1. As the vertical angles are to be used to determine height differences, the heights above ground of the theodolite and targets must be known (Fig. 2.50).

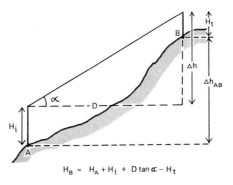

$$H_B = H_A + H_i + D \tan \alpha - H_t$$

Fig. 2.50 Trigonometric heighting.

2. The vertical circle must be levelled by centring the vertical circle index bubble. This adjustment must be carried out before every vertical reading and can be considered similar to the levelling of the main bubble in a tilting level (see sect. 2.3.4.1).

3. A check should be made to determine the readings for angles of elevation and depression. This can be done by pointing the telescope well above the horizontal, noting the reading, and then bringing the telescope through the horizontal to a depression pointing and noting the way in which the vertical circle reading changes. This procedure is normally carried out on face left (Fig. 2.51). A sketch should be made on the booking sheet to record the values for angles of elevation and depression to avoid confusion with small vertical angles.

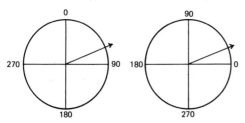

Angle of elevation : Reading 80 °   Angle of elevation : Reading 10 °

Fig. 2.51 Two examples of vertical circle orientation.

The steps required for the measurement of a vertical angle are as follows.

1. Sight the target.
2. Level the vertical circle index bubble.
3. Take the reading.
4. Note the point on the target to which the reading has been taken.

2.3.2.2 BOOKING OF VERTICAL ANGLE OBSERVATIONS

One method of booking the vertical circle readings and computing vertical angles is shown in Table 2.2. This format is similar to the one given for horizontal angles, the main difference being the need to compute the vertical angle on FL and FR before taking the mean. This is necessary because a change in face does not bring a change of approximately 180° in the vertical angle as it did with horizontal angles. The reason for this is illustrated in Fig. 2.52.

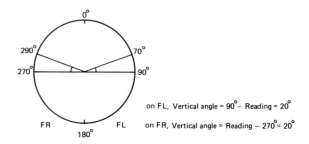

on FL, Vertical angle = 90° − Reading = 20°

on FR, Vertical angle = Reading − 270° = 20°

Fig. 2.52 Computation of vertical angles.

It is not necessary to record vertical angles in any specific order, as each individual observation is taken with respect to a level line defined by the vertical circle index bubble. Vertical angles can be recorded at the same time as horizontal angles, but if measured on their own there is no need to complete FL readings before starting on FR, nor must a strict order be followed in the observations. If multiple observations are taken, there should be the same number of FL and FR pointings used to determine the final mean vertical angle.

Table 2.2 Booking sheet for vertical angles

Station	To	Face	Reading	Vertical angle	Mean	Description
Park	Spire	FL	86 24 10	+03 35 50	+03 35 45	
		FR	273 35 40	+03 35 40		
	Pole	FL	92 10 05	−02 10 05	−02 10 15	
		FR	267 49 35	−02 10 25		
	Tree	FL	89 51 40	+00 08 20	+00 08 15	
		FR	270 08 10	+00 08 10		

2.3.2.3 CORRECTIONS FOR CURVATURE AND REFRACTION EFFECTS

Since the earth is almost spherical, the difference between a horizontal line and a level line at a point must be investigated. A horizontal line is defined as one at right angles to gravity at a point, whereas a level line is defined as a line of equal height above some datum.

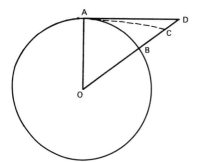

Fig. 2.53   Earth curvature and refraction.

In Fig. 2.53, AD is the horizontal line at A while AB is the level line. As the length of AB increases, so will the discrepancy DB, which is the vertical distance at point B between the level line and the horizontal line from A. This is the effect of earth curvature.

If a measuring staff was set up at B in the position BD (i.e. at right angles to the horizontal line at B) and was viewed along the horizontal line from A, point C on the staff would be observed, not point D. This is because the light ray travelling from C, where the air is less dense, to A will tend to curve towards the vertical as it passes through layers of air of increasing density. This is the effect of atmospheric refraction.

The total or combined effect of curvature and refraction has been computed as being

$$\Delta h(\text{ft}) = 0.57 \, s^2, \text{ where s is in miles}$$

or

$$\Delta h(\text{m}) = 0.0675 \, s^2, \text{ where s is in km.}$$

This correction must be added to the height difference computed from the formula $\Delta h = D \tan \alpha$. Over small distances this correction is negligible, but over longer distances it can be very significant. For example

for s = 1 km, Δh = 0.068 m,
for s = 10 km, Δh = 6.750 m.

2.3.2.4 APPLICATIONS OF TRIGONOMETRIC HEIGHTING

Trigonometric heighting, using a theodolite to measure the vertical angles, is the standard method of carrying height through a triangulation scheme or along a traverse, where the horizontal angles have been observed using a theodolite. For all but the shortest of lines, the vertical angles must be measured to a high accuracy and the effects of curvature and refraction taken into account. The formulae given above for the curvature and refraction corrections are approximate, and more sophisticated procedures must be adopted if high accuracy is required over long distances.

This indirect method of height difference determination is of great value in surveys carried out by the radiation method, especially when tacheometry is employed (see sect. 2.4.3) or when using electronic distance measuring equipment (see sect. 2.6.1.3).

2.3.3 BAROMETRIC HEIGHTING

Barometric heighting is another indirect method of determining height differences. The difference in the air pressure at two points is converted into a height difference; the principle of this measurement is shown in Fig. 2.54. Somewhere in the upper atmosphere a plane of constant pressure is assumed, and this plane, known as an isobaric surface, is the base for the measurement of pressure. At point A the pressure will be less than at point B, because there is a smaller column of air

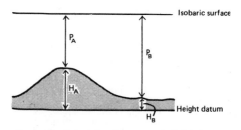

Fig. 2.54   Principle of barometric heighting.

above A than above B. This pressure difference must be related to the difference in height.

2.3.3.1 EQUIPMENT

Barometers used in the field are of the aneroid type. These have a partially evacuated chamber, the sides of which move when air pressure changes. These movements are linked by a gearing system to a dial on which either air pressure or height can be read (Fig. 2.55).

Fig. 2.55   Aneroid barometer.

There are two main problems to be overcome in barometric work. First, the relation between absolute pressure and absolute height is a very complex one and cannot be illustrated by a simple formula. Relating pressure differences to height differences is much simpler, and although for highest accuracy computations are still required, for lower orders of accuracy compensated barometers can be used. The second problem concerns the isobaric surface itself. This surface of equal pressure is not always at the same height and may slope or suffer irregularities. These problems can be overcome by working only in stable meteorological conditions and restricting movement to a few kilometres at a time.

2.3.3.2 FIELD PROCEDURE

There are many different procedures used in barometric heighting and these are described at length in standard survey texts. Only two simple methods will be given here, for use with a compensated barometer which gives readings directly in metres.

With a single barometer, small circuits are run from the base station, returning there after every few readings (Fig. 2.56).

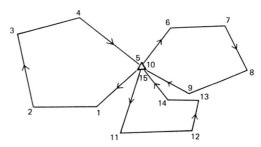

Fig. 2.56   Use of a single barometer.

If the pressure over the survey area is constant, the height readings at the base station will remain constant, but if the pressure changes, the following graphical adjustment can be carried out. The height readings are plotted against time and the plotted positions of the base station are joined so that a datum for height measurement is obtained. This assumes that the change in pressure over the area is uniform and thus that the datum on the graph will be a straight line or a smooth curve (Fig. 2.57). The difference in height between the base station and the individual points can be scaled off the plot.

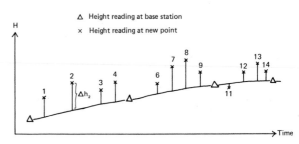

Fig. 2.57   Graphical adjustment of single barometer reading.

A much better procedure uses two barometers, one kept at the base station and read at regular intervals, while the other is moved to the points of interest. This allows the changes in pressure due to

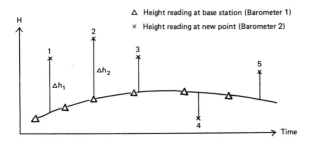

Fig. 2.58 Graphical adjustment using two barometers.

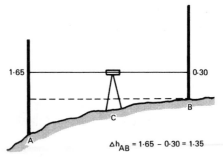

$$\Delta h_{AB} = 1.65 - 0.30 = 1.35$$

Fig. 2.59 Principle of levelling.

weather changes to be recorded more accurately. The reduction is similar to the method previously discussed if the readings on both barometers are made equal at the base station at the start of the survey. However, a more precisely defined datum line is obtained (Fig. 2.58).

2.3.3.3 ACCURACY AND APPLICATION

There are two aspects of accuracy to consider here – that with which the instrument can be read and that of the final heights. With small mountain barometers, readings can be taken to ±3 m, and the final heights will be correct to about ±5 m. Survey barometers used in the correct manner can give final heights correct to ±2–4 m, which is sufficient for many survey purposes.

Barometers are not used much by surveyors, but they can be of great value to field scientists, especially in exploration and reconnaissance mapping when the height accuracies obtainable are more than adequate for many purposes. In vegetation studies, the upper limit of some vegetation types can be obtained to a higher accuracy than by estimating one's position on a small-scale and often unreliable contour map. The approximate altitude of a col or a pass in geological or geomorphological studies, the height of head dykes, former limits of cultivation and upland settlements can all be determined well enough for most purposes with a barometer, using a simple procedure.

2.3.4 LEVELLING

Levelling, which is the most accurate method of heighting, is based on the direct measurement of

the height difference between a pair of points. The procedure is to set up a horizontal line of sight (or line of collimation) using a surveyor's level, and to read the point at which this line of sight cuts a graduated staff held vertically over the point of interest.

With a level set up at point C, the readings to the staff at points A and B might be 1.65 m and 0.30 m (Fig. 2.59). As the staff is graduated from the bottom, these readings indicate the vertical distances from the ground to the horizontal plane defined by the level. At A, it is 1.65 m to the ground and at B only 0.30 m, so that the difference in height between A and B will be

$$1.65 - 0.30 = 1.35 \text{ m}.$$

The situation is more complicated than is shown in Fig. 2.59, for the earth is not flat. This problem was introduced in section 2.3.2.3. In Fig. 2.60, the horizontal line CZ, as defined by the level, is at right angles to the vertical at the instrument; the level line BY, however, is curved and at a constant distance from the centre of the earth. The difference in height between two points is then more accurately defined as the vertical distance between two level lines passing through the two points, e.g. AX' or A'X. Further, if the height above datum $\bar{A}A$ of A is known, the absolute height of X can be computed.

If the simplified situation of Fig. 2.59 is related to the actual situation shown in Fig. 2.60, it can be seen that a levelling procedure based on horizontal instead of level lines of sight can be developed if one of two conditions holds.

1. The level is midway between A and X, (CB = ZY, and therefore BA − YX = CA − ZX).

Fig. 2.60 Basic levelling 'rectangle'.

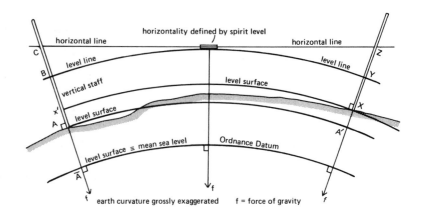

horizontality defined by spirit level

horizontal line
horizontal line

level line
level line

vertical staff
level surface

level surface

level surface ≡ mean sea level
Ordnance Datum

earth curvature grossly exaggerated   f = force of gravity

2. The distance from the instrument to the staff is so short (say less than 150 m) that the level line and the horizontal line can be considered to be coincident.

Usually, more than one set-up of the instrument is necessary in order to join points of interest, and basic levelling rectangles (as defined in Fig. 2.59) must be linked in a series. This line of levels, or series levelling, is therefore a series of leapfrog repetitions of the basic rectangle, linked together at the change points, where the staff is read twice, from two instrument positions. The first reading from any instrument station is called a backsight (BS), the last a foresight (FS), and any other reading is called an intermediate sight (IS). Figure 2.61 illustrates the principle of series levelling, including the use of intermediate sights; the field procedure will be discussed later (sect. 2.3.4.3).

2.3.4.1 EQUIPMENT

Levelling requires an instrument called a level, which defines a horizontal line, and a staff graduated in suitable units. Although any instrument which can define a horizontal line of sight can be used for levelling, e.g. the hand-level (Fig. 2.62) or a theodolite, the term level is usually restricted to instruments specifically designed for this work.

Modern levels achieve stability by their tripod mounting, magnification and good resolution by their optics, and consistent definition of the horizontal plane by spirit levels and pendulum devices of varying degrees of accuracy and sensitivity. Many variations in cost, optical performance, appearance and accuracy are available; some have refinements for more accurate reading of the staff and some embody devices that will accomplish additional survey operations. In essence, however,

Fig. 2.61 Series levelling.   A LINE OF LEVELS (Earth curvature ignored)

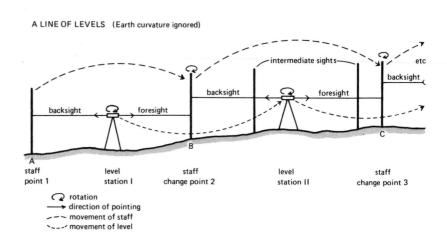

intermediate sights   etc

backsight

backsight   foresight

backsight   foresight

staff
point 1

level
station I

staff
change point 2

level
station II

staff
change point 3

↻ rotation
→ direction of pointing
--- movement of staff
----- movement of level

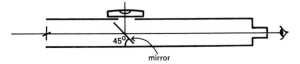

Fig. 2.62 Hand level.

every level is designed to produce a line of sight or collimation that consistently and accurately defines the horizontal. There are three main types of instrument: the Dumpy level; the tilting level; and the automatic level.

Several types of mounting are used to attach the instrument to its base plate, which in turn is fixed to the tripod by the fastening clamp (Fig. 2.63). Footscrews or ball and socket mounts can be manipulated to level the instrument. Spirit levels, attached to the instrument in general and the telescope tube in particular, indicate when the

instrument is in the position required for a horizontal line of collimation.

Three-footscrew levelling has already been described (see Fig. 2.44) for the theodolite, and a similar procedure would be used for a Dumpy level. For a tilting or automatic level, the procedure is simplified by the use of a small bull's-eye bubble. With the telescope set (for convenience) parallel to any two footscrews, as shown in step one in Fig. 2.64, the bubble is centred using the three footscrews or the ball and socket head, with no need to rotate the telescope during these operations.

Final precise levelling of the instrument is achieved with the tilting screw (Fig. 2.64), after the circular bubble has been centred and after the telescope has been pointed at the staff. The tilting screw tilts or pivots the telescope tube, the more accurate tilting bubble showing when the line of collimation is horizontal. An inclined mirror or

Fig. 2.63 Components of a simple tilting level.

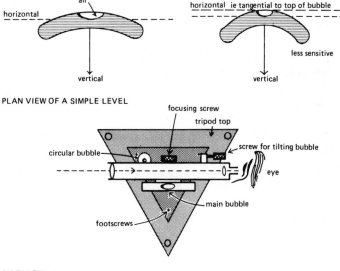

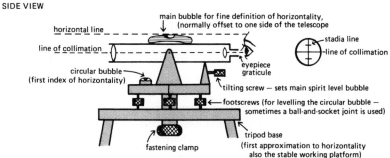

Fig. 2.64 Levelling a tilting level.

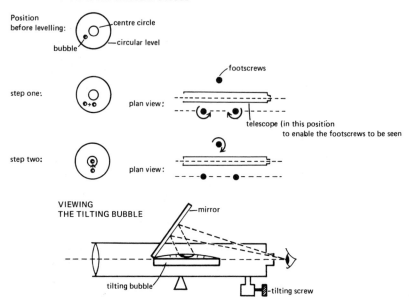

peep-hole is provided close to the eyepiece of the telescope so that the observer need only shift his viewpoint slightly to make and check this critical setting.

In an automatic level, there is no main bubble, only the small circular bubble which is used to bring the instrument into an approximate horizontal position. Once the circular bubble is centred, an automatic system within the instrument comes into operation, ensuring that what is seen against the crosshairs is always a horizontal line of sight. This is achieved by a system of prisms, one of which is suspended or balanced on a fulcrum edge in such a way that when it comes to rest under the influence of gravity, the light rays are refracted at its surfaces to bring the horizontal ray from the object being viewed on to the crosshairs of the diaphragm. For any position of the level, when the circular bubble is within its circle, the line of sight will be horizontal. Although more expensive than tilting levels, automatic levels are very useful when many observations have to be taken, and the extra cost will be more than offset by the savings in time and effort.

Levelling staffs are of two types: telescopic (normally three or four sections) and folding, the latter being hinged or detachable. Many forms of

graduation are available and the more common metric types are shown in Fig. 2.65. Mistakes are often made if staff graduations are considered to be equivalent to scale divisions on a ruler. On a surveying staff, the black line is a unit and not merely an index line.

Particular care must be taken near the top of each 10 cm section, as it is common for gross errors to be made in this region. Familiarity with reading the staff is necessary before any survey is attempted, and practice can be obtained in the office by using a pointer and staff section, turned upside down if the level to be used gives an inverted image.

The staff should be held vertically, and many staffs incorporate a small circular bubble to indicate when the staff is being held correctly. If no circular bubble is available, the staff holder must do his best to maintain the correct position. The vertical hair of the telescope graticule will detect any sideways lean, which can be indicated to the staff holder by appropriate hand signals. A lean towards or away from the instrument is more difficult to detect. If the staff is swayed gently backwards and forwards, the lowest reading will give the correct result.

A useful device, placed on the ground, and on to which the staff can be placed, is a small wooden

Fig. 2.65 Staff readings.

GRADUATIONS ON STAFFS

READING THE STAFF

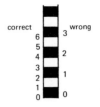

THE STAFF VIEWED THROUGH THE EYEPIECE

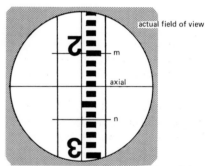

the staff (½cm)

actual field of view

m

axial

n

care must be taken to read the central (axial) line on the graticule
and not the shorter stadia lines

or metal plate which ensures that the staff does not sink into the ground between readings and is of particular value at change points, when the direction of the staff has to be altered.

For normal levelling, the survey team consists of an observer and booker at the instrument and a staff holder, although with experience the observer may wish to do his own booking. If many intermediate sights are to be taken, or if the terrain is difficult to cross, two staffs may be used, observed alternately.

The observer may be assisted by a booker and staff holder who are not surveyors, but it is worthwhile for the team to discuss the problems and route fully before the survey begins, in order to eliminate undue delay during the work.

2.3.4.2 CHECKING A LEVEL

Before using a surveyor's level, a check must be carried out on the instrument. Different instruments require different procedures and the manufacturer's handbook should always be consulted. The adjustment most likely to be needed is one to eliminate collimation error, which occurs when the line of sight as indicated by the crosshairs is not parallel to the horizontal line as defined by the bubble. This error can be detected and eliminated by the method shown in Fig. 2.66. As this error has no effect if backsights and foresights from the same instrument station are equal, an adjustment is not always necessary, but if intermediate sights are to be used or if the terrain would make equal sights

Fig. 2.66 Collimation error.

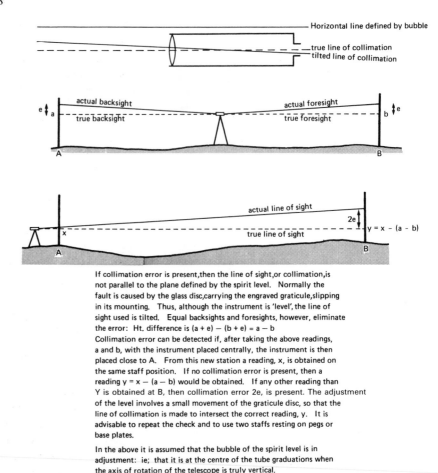

If collimation error is present, then the line of sight, or collimation, is not parallel to the plane defined by the spirit level. Normally the fault is caused by the glass disc, carrying the engraved graticule, slipping in its mounting. Thus, although the instrument is 'level', the line of sight used is tilted. Equal backsights and foresights, however, eliminate the error: Ht. difference is $(a + e) - (b + e) = a - b$

Collimation error can be detected if, after taking the above readings, a and b, with the instrument placed centrally, the instrument is then placed close to A. From this new station a reading, x, is obtained on the same staff position. If no collimation error is present, then a reading $y = x - (a - b)$ would be obtained. If any other reading than Y is obtained at B, then collimation error 2e, is present. The adjustment of the level involves a small movement of the graticule disc, so that the line of collimation is made to intersect the correct reading, y. It is advisable to repeat the check and to use two staffs resting on pegs or base plates.

In the above it is assumed that the bubble of the spirit level is in adjustment: ie; that it is at the centre of the tube graduations when the axis of rotation of the telescope is truly vertical.

difficult to arrange, the adjustment should be carried out.

2.3.4.3 FIELD PROCEDURES

As with planimetric surveying, a distinction can be made between control levelling and detail levelling, the former being the operation for the determination of heights of points to be used as control for the detail levelling that follows.

Every survey should begin with a reconnaissance, either in the field or using stereo aerial photographs (see Ch. 3), in order to minimise the number of instrument stations and to avoid steep ground and marshy areas. A field sketch, showing the main features of the area with respect to the objectives of the survey, can be made during the reconnaissance and this will be used for route plan-

ning. A straight line may not always be the quickest or easiest route between two points; a line following a gentle gradient and avoiding broken ground will normally be better, allowing equal backsights and foresights.

The control for levelling is provided by Ordnance Survey bench marks (Fig. 2.67). A network of these points is found throughout Great Britain and their precise heights are referred to the National Datum or Ordnance Datum (OD) of mean sea level, as defined by tide gauge readings taken over a period of six years (1915–21) at Newlyn in Cornwall. (The Irish Datum refers to mean sea level at Dublin, and in some islands around the coast of Britain a local datum may be used.) Bench marks are systematically relevelled by the Ordnance Survey and the heights are located and supplied on the most recent 1:1250 and 1:2500 scale plans. Another source is the Bench Mark List

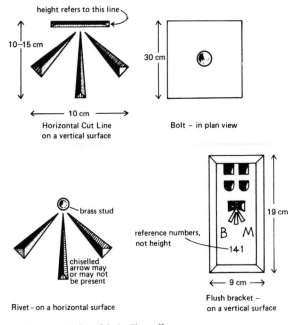

height refers to this line

10–15 cm

← 10 cm →

Horizontal Cut Line
on a vertical surface

30 cm

Bolt – in plan view

brass stud

chiselled
arrow may
or may not
be present

Rivet – on a horizontal surface

reference numbers,
not height

B M

14·1

19 cm

← 9 cm →

Flush bracket –
on a vertical surface

In all instances the foot of the levelling staff
should be rested on the point shown by the
arrow in the above diagrams. The exception
is a pivot bench mark which is a small hollow
cut in a horizontal surface and should be used
with a 16mm diameter ball bearing or glass marble
as the staff support.

Fig. 2.67 Ordnance Survey bench marks.

for the relevant National Grid kilometre square, obtainable from the Ordnance Survey (address in Appendix 1). All bench marks have been heighted to an accuracy greater than required for topographic mapping purposes and are quoted to 0.01 ft and 0.01 m. Thus a typical bench mark description on a list might be:

New bench mark at wall junction with fence SE side road

NG 10 m ref. 7552 4637
Altitude 9.07 ft.   Date of levelling 1955
          2.77 m

More information on the O.S. products is given in Appendix 1.

The traditional arrow-shaped groove chiselled on a vertical surface is not the only type of bench mark. Flush brackets, bolt bench marks, rivet bench marks and pivot bench marks are all used and are sometimes rather inconspicuous. A surprising amount of time may be lost searching for bench marks in the field, even with the aid of a

large-scale map and it is strongly recommended that during the course of the preliminary reconnaissance all available bench marks should be visited and their positions recorded by point sketches. Figure 2.67 shows a selection of the bench marks used in Great Britain. The use of O.S. bench marks allows absolute heighting, which is needed if results gained are to be compared with those from other sites or features elsewhere in the country. The value of relative, or local heights, is limited to the area of the survey, but are sufficient to work out a gradient. Equally, in the study of river terraces, if it is sufficient to know that terrace B is X metres above terrace A, then the difficulty and extra effort required to produce absolute heights related to the O.S. datum is unnecessary. However, if in a botanical study, the maximum altitude of growth of a specific plant assemblage is required in different parts of the country, all measured heights must refer to a single datum and that should be the O.S. datum.

The use of bench marks has a further advantage. As all bench marks have been surveyed to the same high accuracy, it is possible to begin levelling on one bench mark and end on another, which may be more convenient than the alternative of returning to the starting point.

To illustrate the field procedure, and the booking and reduction of the data, a simple example is given. The levelling task can be divided into two parts.

1. A level line is run from an O.S. bench mark into the area, establishing several TBMs (temporary bench marks), and the line is checked by closing on another O.S. bench mark.
2. Using the TBMs, the heights of the points of detail within the survey area are established.
     This situation is shown in Fig. 2.68.

Booking and reduction of the observations are carried out on a special form. For line levelling, where there are few intermediate sights, the rise and fall method is recommended. The readings and reductions are shown in Table 2.3.

The readings are recorded and the computations carried out in the following way.

1. Each point on the ground where the staff is placed is represented by a horizontal line on the form, and as the staff is read twice at a change

Fig. 2.68 Levelling project: booking and reduction of the observations.

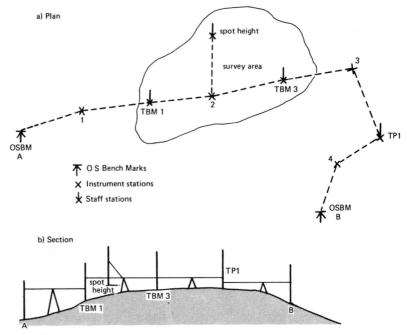

a) Plan

spot height

survey area

TBM 3

TBM 1

OSBM A

3

2

1

TP1

4

OSBM B

⊤ O S Bench Marks
✗ Instrument stations
✶ Staff stations

b) Section

spot height

TP1

TBM 3

TBM 1

A

B

point, both the backsight and the foresight to this point must be on the same line.

2. To obtain the rise or fall, each I.S. or F.S. is compared to the previous I.S. or B.S., depending on the entry which appears on the previous line, e.g. with reference to Table 2.3, the difference in height between OSBM A and TBM 1 is 1.785 and this is a rise.

3. No comparison is made between readings entered on the same horizontal line.

4. To check the arithmetic, the following conditions should hold:

(sum of B.S. − sum of F.S.)
= (sum of rises − sum of falls)
= (last reduced level − first reduced level).

Table 2.3 Booking sheet for levelling: the rise and fall method

B.S.	I.S.	F.S.	Rise	Fall	Reduced level (m)	Remarks	Adjusted values (m)
2.650					24.690	OSBM A	
2.010		0.865	1.785		26.475	TBM 1	26.485
	1.865		0.145		26.620	SPOT HT	26.637
0.685	0.220	1.645			28.265	TBM 3	28.287
1.010		2.155		1.470	26.795	TP 1	
		2.025		1.015	25.780	OSBM B	25.820
Σ6.355 5.265	Σ5.265	Σ3.575 2.485	Σ2.485				
1.090			1.090			1.090	

Fig. 2.69   Adjustment of level line.

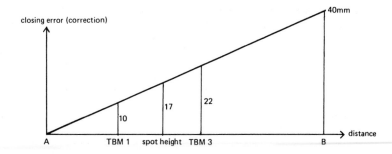

So far, no use has been made of the second OSBM at B. This point can, however, give a check on the field work as its computed height can be compared with its known height. The closing error is defined as the difference between the known and computed value of the final point, which is 0.040 in the worked example. This discrepancy may be ignored if it lies within the tolerance of the survey; if it is very large it indicates a gross error and the field work may have to be repeated. When the error is too large to ignore and too small to indicate a gross error, it can be distributed along the line of levels. This can be carried out graphically, as shown in Fig. 2.69, with the adjusted values written in the last column of the table.

With the main level line closed, work can begin on the heighting of points of detail. An instrument station is selected such that a backsight to a TBM is possible and a good view of the points to be heighted is obtained (Fig. 2.70).

With the level set up at point F, a backsight is taken to TBM 1, say 2.465. As the height of TBM 1 is 26.485, the height of collimation of the instrument will be

$$26.485 + 2.465 = 28.950.$$

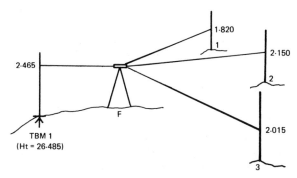

Fig. 2.70   Use of a TBM for heighting points of detail.

If an I.S. is taken to point 1, the height of the point can be obtained simply by subtracting the reading from the height of collimation. When all possible readings have been taken from this instrument station, an F.S. is taken to a change point; a B.S. to this change point from the new instrument station gives the new height of collimation, and the procedure continues as before. This type of booking is called the height of collimation method and is shown in Table 2.4. A simple check is available for the change points (see Table 2.4); more complicated checks are available for the I.S.s but recalculation, by a different person if possible, is usually more efficient.

As these checks only apply to the arithmetic and not to the observations, this line would have to finish on a known point and adjusted if necessary.

A line of levels may begin on a point given an arbitrary value, defining a purely local datum, which may be sufficient if only local height differences are required. If this point is permanently marked, any future level line run from it to an Ordnance Survey bench mark would allow the local heights to be transformed to the O.S. datum.

In levelling, the proper use of bench marks satisfies the two rules of working from the whole to the part and incorporating independent checks. A line of levels is therefore only half completed when it reaches the point for which height is required. If levelling has to stop, due to bad weather or any other reason, without having closed on to a known point, then the heights so far obtained are unreliable since there is no check as to whether or not an error has been made. A F.S. should be taken as the last observation on to a TBM, from which the survey can restart and proceed to a known closing point.

For the field scientist who is working in mountainous or remote terrain, where there is scant

Table 2.4 Booking sheet for levelling: the height of collimation method

B.S.	I.S.	F.S.	Height of collimation	Red. level	Remarks
02.465			28.950	26.485	TBM 1
	1.820			27.130	Spot Ht
	2.150			26.800	Spot Ht
	2.015			26.935	Spot Ht
	0.325			28.625	Spot Ht
1.655		3.400	27.205	25.550	Change Point
	3.225			23.980	Spot Ht
	4.010			23.195	Spot Ht
		2.150		25.055	Change Point
$\Sigma = 4.120$		$\Sigma = 5.550$		26.485	
5.550					
$\Delta = 1.430$				$\Delta = 1.430$	

provision of bench marks, the problem of reducing height values to the absolute datum can be almost insurmountable, unless one is prepared to use some method other than levelling, which is generally a slow and laborious procedure if the distances to be covered are greater than a few kilometres and the total of the rises or falls is more than 40–60 m. The alternatives which can be considered are trigonometric heighting, barometric heighting and the use of parallax differences measured on stereo aerial photographs.

2.4 HOW TO OBTAIN PLAN AND HEIGHT IN ONE OPERATION

In mapping, it is often necessary to obtain both the planimetric position and the height of a number of points, and this can sometimes be done by combining some of the techniques already discussed. For example, level lines could be run through an area already covered by a triangulation and offset survey, heighting points previously fixed in plani-

metry, but this method would be suitable only if a limited number of height points were required.

Sometimes, taping and levelling can be combined with advantage in the running of cross-sections or profiles. Where horizontal distances are known, height differences can be computed if vertical angles are measured. However, for a detailed survey where the three-dimensional coordinates of a large number of points have to be established, the methods previously discussed are not adequate and a tacheometric method should be employed. The use of a theodolite and a plane-table alidade for tacheometry will be discussed after some of the other possibilities have been investigated.

2.4.1 PLANE TABLE AND INDIAN CLINOMETER

The plane-table procedures for intersection, resection, radiation and traversing allow the position of new points to be plotted on the plane-table sheet. The horizontal distance between these new points and previously plotted points can be obtained by scaling off the distance on the plot.

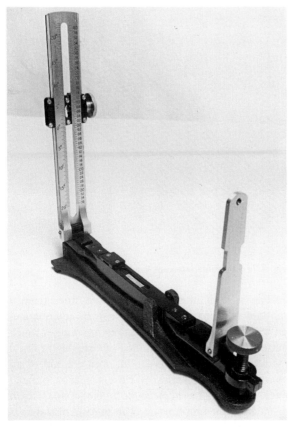

Fig. 2.71   Indian clinometer.

With the horizontal distance known, the difference in height between the two points can be determined if the vertical angle is measured. An Indian clinometer is specially designed to measure these vertical angles (Fig. 2.71).

The instrument consists of two vertical arms mounted at right angles to a plate containing a spirit bubble that can be levelled using a tilting screw. On one arm is a pin hole, and on the other a crosshair movable in the vertical direction. This crosshair is adjusted so that when one looks through the pin-hole, one sees the target to be measured. The vertical arm on which the crosshair slide moves carries two scales, one graduated in degrees and the other in natural tangents. The difference in height is computed from

$$\Delta h = D \tan a,$$

where D is the horizontal distance.

2.4.2 TRIGONOMETRIC HEIGHTING BY THEODOLITE

This method has already been discussed (see sect. 2.3.2). If a numerical triangulation is to be carried out using angles measured by theodolite, then, by taking vertical angles also, the heights of the points can be computed. Over moderate distances, this is the easiest and most accurate method of obtaining height differences, provided that the curvature and refraction correction is applied.

2.4.3 COMBINATION OF LEVELLING WITH DISTANCE MEASUREMENT

One simple method of obtaining planimetric and height information together is by surveying a cross-section or profile. This entails levelling along a line, while noting with each height reading the distance travelled along the line.

A good practical example of this type of work is beach profiling. From a fixed point at the top of the beach, a line is defined by magnetic compass, by two transit marks or by setting off an angle from a known or reference direction using a theodolite. A chain is then laid along this line and levelling is used to height points every few metres along the chain or at significant breaks of slope along the profile. In addition to the staff reading, the chainage distance to each point is noted. The alignment of the chain, the counting of the whole chain lengths and the way of dealing with steep slopes are described in section 2.2.2. In general, no slope correction is necessary, especially if the profile is to be presented graphically. Changes in the beach profile can be detected by levelling along the same lines at regular intervals. To emphasise small changes in elevation, a vertical exaggeration can be applied to the plot.

The same basic idea can be applied to a trilateration and offset survey if heights are also required. By taking staff readings at intervals along the main chain lines, one can define the position of the staff stations by the recorded chainage distances. Chains can also be laid out between other points of known position, and in this way the area can be crossed by lines of points of known position and height.

When more height information is required than can be acquired by joining points of known position, a grid of points can be set out and measured. Assuming that the area below line AB in Fig. 2.72 requires detailed height information, a chain is laid along part of the line and right angles are set at each end of the chain (C and D). Chains are laid along lines CE and DF at right angles to line AB. The levelling staff is then placed at, for example, every 2 m mark along the chain laid out between C and D. The chain is then moved to position C′D′ and the staff read again at 2 m intervals.

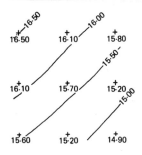

Fig. 2.72 Grid levelling.

In this way, a 2 m grid of points is measured over the area of interest. To plot this information, a grid is drawn up to scale on the line AB, with the base of the grid plotted according to the positions of C and D on the line AB. The height values can then be entered against the corresponding intersection points on the grid. This method is known as grid levelling.

If contour lines are required, they can be obtained by linear interpolation between the grid intersections (Fig. 2.73).

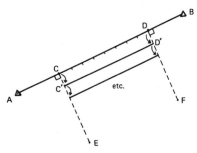

Fig. 2.73 Interpolation of contours on a regular grid of spot heights.

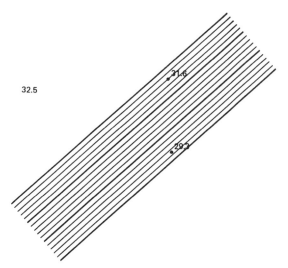

Fig. 2.74 Variable intercept tracing sheet.

The task of interpolating contour lines from a random scatter of spot heights can be tedious and difficult if carried out by visual estimation. A much better solution can be obtained by using a graphical method of linear interpolation, with the aid of an overlay sheet. A series of closely spaced parallel lines is drawn on a piece of tracing paper, with every fifth line emphasised. The spacing of the lines can be used to represent height intervals of 0.1 m, 0.2 m, 0.3 m, etc., depending on the spot-height accuracy. The spacing of the points is accommodated by rotating the tracing paper until the height interval between the spot heights is matched by the height interval given by the lines on the tracing paper. With a ruler laid between the two spot heights, the position of the contour line intercept can be pricked through on to the base map (Fig. 2.74). The variable intercept tracing sheet can then be moved to the next pair of spot heights.

It is usually easier to begin contour interpolation using the highest spot heights in an area (i.e. the summits), working downwards. The contour intercepts between the summit points and a number of lower points should be established before the contour lines are drawn in.

2.4.4 STADIA TACHEOMETRY

Tacheometry is defined as the rapid measurement of distance by optical means, and any instrument

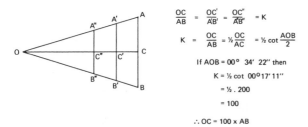

$$\frac{OC}{AB} = \frac{OC'}{A'B'} = \frac{OC''}{A''B''} = K$$

$$K = \frac{OC}{AB} = \frac{1}{2}\frac{OC}{AC} = \frac{1}{2}\cot\frac{AOB}{2}$$

If AOB = 00° 34′ 22″ then

K = ½ cot 00° 17′ 11″

= ½ . 200

= 100

∴ OC = 100 × AB

Fig. 2.75 Principles of stadia tacheometry.

which can measure distance optically is called a tacheometer. Stadia tacheometry is based on the principle that in similar isosceles triangles the ratio between the apex perpendicular and the base is constant (Fig. 2.75).

If A'', C'' and B'' are the stadia hairs on the diaphragm set in such a way that A'' and B'' subtend an angle of 34 minutes and 22 seconds of arc, and if A, B and C are the readings on a staff indicated by each of the stadia lines, then AB subtends the same angle, and the distance from the instrument to the staff will be 100 times the value of the intercept distance AB as read on the staff. If the telescope is horizontal, the height of the point can be obtained by the height of collimation method

(sect. 2.3.4.3), and the direction to the point can be obtained from the horizontal circle reading. With inclined sights, the vertical angle must be recorded in order to reduce the observations. Figure 2.76 illustrates how this is done, but only the final formulae need be remembered.

In tacheometric work, three types of reading are required to fix the position of a point: the horizontal angle, the vertical angle and the three stadia readings. As plotting is carried out using a protractor, it is quite sufficient to read the horizontal angles to the nearest minute of arc and only on one face (usually F.L.). As a check on the stadia readings, the booker should ensure that the middle reading is the mean of the upper and lower values. The staff can be read in three ways, each of which has its own particular advantage.

1. Setting the bottom stadia hair on a whole number on the staff makes the subtraction (which should be carried out in the field as a check on the observation) much easier.
2. Setting the central hair to give a staff reading corresponding to the height of the instrument equalises the values of i and m, and the differ-

Fig. 2.76 Stadia tacheometry with an inclined sight.

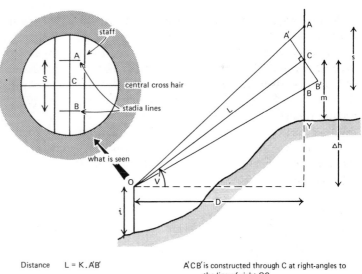

Distance	$L = K . A'B'$
	$= K . s . \cos V$
	$D = L . \cos V$
	$= K . s . \cos^2 V$
Height	$\Delta h = L . \sin V$
	$= K . s . \cos V \sin V$
	$Hy = Hx + i + Ks . \cos V \sin V - m$

$A'CB'$ is constructed through C at right-angles to the line of sight OC.

s = stadia intercept AB
i = height of instrument above ground at X
m = height of C above ground at Y,
  ie reading on staff at C
L = OC = slant distance
D = horizontal distance
Δh = difference in height between O and C
K = instrument constant, usually 100.

57

ences in height between the two stations will be given directly by Δh.

3. Setting the vertical angle to a multiple of 20 minutes of arc allows the use of special tables for the reduction of inclined observations (even angle method).

2.4.4.1 PROCEDURE FOR TACHEOMETRY USING A THEODOLITE

The field procedure for a tacheometric survey is as follows.

1. A control network suitable for a detail survey by the radiation method is first laid out.
2. The theodolite is set up at a control station, the height of the instrument is noted and the direction to another, distant control station (usually called the reference object or RO) is read off the horizontal circle. This value can be pre-set to a specific value, say zero or the bearing of the orienting line, by moving the lower plate of the instrument. This reading is used to orient the horizontal reading for all radiated points.
3. The staff is placed at a new point, sighted and the stadia readings noted. The horizontal and vertical angles can be read while the staffman moves to the next point of interest.
4. Readings are taken to all points of interest, with the horizontal angle to the RO being read from time to time to check that the theodolite has not been disturbed.

Office reductions

The reduction of the observations can be carried out in a number of ways. Tacheometric tables, such as Redmond's tables for the even angle method of reading, are very easy and quick to use, as is a tacheometric slide rule. Although it is possible to perform the reductions using sine and cosine tables, this is extremely tedious and is not recommended. The advent of small electronic calculators, especially those which are programmable, allows the direct reduction of observations taken with any vertical angle, and this is the standard method of reduction now in use.

Some tacheometers are self-reducing in that the horizontal distance and the difference in height can always be obtained directly from the stadia readings without calculation. In other cases, the reductions

can be carried out using readings taken on attachments to the theodolite (e.g. Ewing stadia-altimeter) or plane-table alidade (Beaman arc). Readers are directed to the manufacturers' publications for details of these instruments. The remainder of this section deals only with the standard method of stadia tacheometry.

Once the reductions are complete, the heights and the bearings and distances to all points from the instrument station will be known. A large circular protractor is placed over the instrument station on the plotting sheet and oriented until the line to the RO cuts the protractor at the horizontal reading recorded for the RO. With the protractor fixed to the sheet, the directions to all the other points can be marked at the edge of the protractor against their horizontal readings. The protractor is removed and a scale is used to plot the distances to the detail points along the corresponding directions. The height value can be written beside each point. Table 2.5 shows an example of tacheometric observations and reductions.

Tacheometry is an extremely versatile survey technique. Stadia tacheometry gives an accuracy in distance of between 1/500 and 1/1000, depending on the length of the sights and the ability of the observer to read the fine graduations on the staff. With a height accuracy of between 40 and 60 mm, maps at scales of 1:500 to 1:5000, with contour intervals ranging from 0.5 m to 5 m can be produced by this method.

2.4.4.2 TACHEOMETRIC ALIDADE USED WITH A PLANE TABLE

Stadia tacheometry can also be applied in plane-table work by making use of a tacheometric alidade, which consists of a telescope mounted on a plotting rule. The distance and difference in height between the plane-table station and the new point are found as for theodolite tacheometry, but the alignment of the instrument on the plane table at the time of observation gives directly the direction to the new point. In most instruments, a parallel guidance mechanism joins the plotting rule to the instrument. The great advantage of the plane-table method is that the map is being produced in the field and it is possible to judge, while the work proceeds, whether too many or too few sights are being taken.

Table 2.5 Booking sheet for tacheometric observations

At station: C
Instrument: V-22
Observer: RW
Date: 3.10.86
Time: 09.30 am

Height of station: 10.20 m
Height of instrument: 1.32 m (above ground)
Height of collimation: 11.52 m
Booker: MW

TACHEOMETRY   PLACE   Tarradale   DATE 3.10.86

Station	instr. ht. (i)	Points	Angle observed			Reading of stadia wires (lower)** (upper)	Stadia difference (S)	Axial (m)	Horizontal distance D	Height* difference (Δh)	Difference		Height above Datum		Notes
			Horizontal	Vertical	***						Rise ($\Delta h - m$)	Fall ($m + \Delta h$) or ($m.\Delta h$)	of the instr.	of the point	
C	1.32	RO	176° 21′										11.52	10.20	Station A
		1	277 27	97° 00′		3.283 2.931	0.352	3.106	34.68	4.258		7.36		4.16	wall
		2	89 50	90 00		1.752 1.588	0.164	1.670	16.40	–		1.67		9.85	base of slope
		3	187 38	86 40		1.145 0.685	0.460	0.915	45.84	2.670	1.75			13.27	ridge
		4	169 08	88 00		1.423 1.140	0.283	1.280	28.27	0.987		0.29		11.23	fence

*   height difference from theodolite centre to axial staff reading. Height of ground point in relation to instrument centre is obtained from "Rise" or "Fall" column.

**   these readings are from an "inverted view" telescope, so lower stadia wire gives the higher reading.

*** These angles are for a V-circle with 0° vertically up (zenith). Examples are given of the four possibilities: (1) a depression angle (always a "Fall"); (2) a 90° level sight (always a "Fall" equal to axial reading); (3) an elevation angle with a "Rise" (usual case) and (4) an elevation angle with a "Fall' (unusual).

2.4.4.3 PROFILING BY TACHEOMETRY

Tacheometry can also be used to survey profiles. Several ranging poles can be set out to define the line along which measurements are to be taken, and the staff holder selects significant features on that line. A check on the field work is available if the cross-section is plotted out in plan, when all the points should plot out in a straight line.

Many modern levels have a stadia diaphragm fitted as standard. For reasonably flat terrain, it is possible to use a level as a tacheometer if the level has a horizontal circle to provide the horizontal direction of each sight. This procedure offers many advantages: the readings and settings are fewer because of the constant horizontal sight, there is no difficulty in reducing the observations and mistakes are less frequent. Heighting accuracy is as good as that obtained in levelling. If the field mapping problem is concerned with micro-changes in height that may be critical, e.g. in salt marsh vegetation studies, a level is superior to a theodolite. The use of a level for tacheometric profiling is not recommended for uneven terrain.

2.4.4.4. LOCATION OF CONTOURS

If the terrain is very flat, or if contours are required to a high degree of accuracy, the interpolation of contours from an array of spot heights would result in a great number of spot heights being required, many of which would not contribute to the position of the contours. A better solution would be to locate the contours in the field, and this can be accomplished by a combination of tacheometric and levelling procedures.

1. A level equipped with a stadia diaphragm and a horizontal circle is set up over a point of known position and height. The horizontal circle reading to a second point of known position is taken to orient the later horizontal circle readings. (These operations are similar to those for setting up a theodolite for tacheometric operations.)
2. The height of collimation of the level is computed using the station height and the instrument height, or a backsight on to another known height point.

3. Using the collimation method of level reduction, it is now possible to compute the staff reading required to give a contour point:

e.g. Height of station   = 140.3 m
      Height of instrument =   1.2 m
      Height of collimation = 141.5 m

Therefore, for the 140 m contour, a staff reading of 1.5 m is required.

4. The staffman is directed by the observer so that the middle hair reading corresponding to the contour value is obtained.

5. The horizontal circle and the stadia readings are recorded, allowing the plotting of this contour point.

6. Once a contour has been found, it is easier to find the next contour point. The staffman attempts to move along the contour, the corrections indicated by the observer reducing with experience.

2.5 SURVEYING UNDERWATER FEATURES

Surveying underwater features is often considered quite separately from land surveying. While some aspects of hydrographic surveying are undoubtedly far removed from normal land surveying operations, the basic requirement of each task is much the same. Hydrographic surveying has to cope with the normal topographic problem of establishing three-dimensional coordinates of a large number of points; the planimetric coordinates define the position of a point with a known height or depth. The basic difference between hydrographic and land surveying is that in the former the third dimension is measured below a certain datum, usually a water surface, and in the latter it is measured above a defined surface, again very often a water surface (e.g. mean sea-level).

The depth of water can be established in a number of ways. In shallow water, a ranging pole or levelling staff may be used, and if the bed is soft, a plate can be fixed to the base of the measuring rod to prevent it sinking during the measurement. If the current is strong, it may be difficult to hold the pole vertical. A second approach is to use a sounding line, or lead line, which is simply a heavy lead weight on the end of a strong non-elastic line that can be marked off in fathoms, feet or metres. The lead line is suitable for water deeper than can be attempted with a measuring rod. Once again, care must be taken that the line is vertical and taut at the moment of measurement. The third and most sophisticated method of determining depth is by the use of an echo sounder, a device emitting sonic pulses that are reflected from the bed of the lake or sea back to the instrument. From the time taken to travel this double distance, and the velocity of these signals in water, the depth can be calculated. In most instruments, depth is displayed on a dial or on a paper trace, so that the surveyor has the result immediately.

Fig. 2.77 Using a fathometer on Loch Ness.

Modern hydrographic surveys, especially in the open sea, are increasingly dependent on sophisticated electronic systems. Satellite navigation methods are used in conjunction with shore-based systems such as Decca Hi-Fix, etc., both to navigate the survey vessel out of sight of land and to fix the position of each echo-sounder reading. While many of these systems are becoming ever more simple to use, their cost puts them well beyond the means of most field scientists, who must seek more conventional and less expensive solutions, some of which will now be discussed.

2.5.1 RIVERS

In most river investigations, the survey must produce data on cross-sections and on longitudinal profiles. If the river is shallow enough to allow wading, a chain can be laid across or down the bed of the river and a staff can be placed at regular intervals along the chain. The depth of water can be read off by the staff holder; the level of the river bottom can be obtained by sighting the staff with a level from the bank, after taking a backsight to a suitable point of known height. If a sufficient section of staff protrudes above the water level, tacheometry by theodolite is preferable and quicker for both longitudinal and cross profiles.

If wading is not possible, a boat must be used. A tight wire can be stretched over the river, with distances marked off at the required intervals. When the boat is at the required position, a depth reading is taken by staff, lead line or echo sounder.

In practice, hydrologists prefer to use bridges, weirs or fixed wires with pulleys for raising and lowering instruments into the river. Any method which avoids being in or on the river is preferred, especially if systematic readings are required over a long period of time. Wading and boats may be practicable at low stages and low discharges, but during flood conditions the same section may be quite impossible.

By dropping a lead line from a bridge, it is possible to obtain a bottom profile of a river bed (see Fig. 2.78). If the current is measured at representative points within each section of the profile (three in the example illustrated), the discharge of the river can be computed.

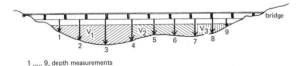

1 ..... 9, depth measurements

V_1, V_2 and V_3, velocity measurements for selected sectors as indicated above.

Discharge is obtained by multiplying velocity with the appropriate cross sectional area. Area usually measured using squared paper.

Fig. 2.78 River discharge measurement.

The survey of large rivers and estuaries, which is beyond the scope of this book, requires considerable expertise and special equipment, as well as substantial logistical support.

2.5.2 SMALL LAKES

In order to obtain depth contours of the bottom of a small lake, a traverse could be run around the lake, establishing the position of points A, B, etc. (Fig. 2.79). A labelled wire is then fixed between pairs of points and depth values are recorded at suitable intervals along the wire. The number of intersecting profiles measured will depend on the density of points required to give a fair representation of the lake bottom.

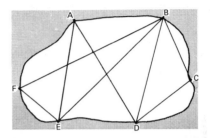

Fig. 2.79 Hydrographic survey of a small lake.

As it is always best to have the profiles running at right angles to the submarine contours, i.e. with the slope, it may be useful to have profiles run parallel to each other. This system also has the advantage of giving a more systematic coverage of the lake bed, with fewer gaps, but extra preliminary survey work is necessary.

If A and B are two traverse points, where the banks are more or less parallel (Fig. 2.80), base lines can be laid off at right angles to the line AB at A and B, and points marked, say, every 15 m along these lines, e.g. A_1, A_2, etc. and B_1, B_2, etc. The wire can then be moved from AB to A_1B_1, then to A_2B_2, and so on, and a grid of depth values can be built up as was done in grid levelling (sect. 2.4.3).

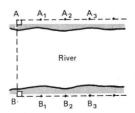

Fig. 2.80 Running parallel sounding lines.

2.5.3 LARGE LAKES AND OPEN SEA

In larger inland bodies of water, the distances are usually too large for the wire techniques to be employed, and in the open sea these cannot be used, but depth recording can be carried out by one of the methods already discussed.

It is useful to make a distinction between the survey operations required to fix the planimetric position of the boat at the moment of sounding and those required to keep the boat on a straight line. The survey boat can be kept on a straight course using a compass, a theodolite or transit marks on the shore. With the two instrumental methods, signals have to be given to the boat to correct its course, but with transits the helmsman can see for himself if he is on line (Fig. 2.81).

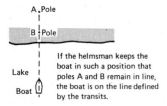

Fig. 2.81 Use of transit marks.

The fixing of position, say every minute, can be carried out by a variety of methods, including radiation, intersection and resection. If the shore station keeping the boat on line is also equipped with a tacheometer or rangefinder, the position of the boat can be determined by measuring the distance to it at any moment, providing that it stays on line. A separate station for fixing the position of the boat can be used for many lines. In this case, both the angle and the distance to the boat must be measured.

Intersection can also be employed. A fixed point on the boat, such as the mast, must be observed by two theodolite observers simultaneously. If the theodolite stations are selected such that the inter-section angle is good, a strong fix will be achieved. There are difficulties, however, in ensuring that the two theodolite observations are taken at exactly the same moment and a suitable signalling system is necessary for good results.

If two angles are measured which are subtended at the boat by three known points ashore, an instrument known as a station pointer is used in the plotting. This method is similar to the mechanical method of plane-table resection. (2.2.3.2).

If the boat is travelling at a steady speed on a straight course, depth readings can be taken more frequently than position determinations and the profile between the fixes can be found by interpolation. When a trace echo sounder, which continuously records the depth, is being used, the trace is marked at every fix and interpolation between these marks will give the depth between each fix. Corrections may have to be applied to take account of the changing level of the sea with tidal movements. Time of observation is thus an essential additional reading to be taken in marine hydrographic work.

There are no simple solutions to the hydrographic survey of a large area; these areas must be left for the specialist hydrographic surveyor who can call on and use the many electronic systems now available.

2.6 HOW TO CONSTRUCT A CONTROL FRAMEWORK

It has already been shown that in surveys of any size a control framework is necessary to provide a framework on which the detailed survey work can be based and that this framework has to be surveyed to a higher accuracy than is required for detailed work. Sometimes, the same equipment can be used for both types of work, only the procedures in the field being different, e.g. double measurement of main lines in chaining (followed by offsets), or a closed series level line (followed by many intermediate sights).

In many projects, however, a higher accuracy is required in the control survey than can be achieved by the methods so far described. This is especially true when coordinate values are required for the control points, and graphical solutions give way to numerical methods. A brief summary of a few more sophisticated measuring devices will be followed by an introduction to computational survey methods.

2.6.1 SOME EQUIPMENT FOR CONTROL SURVEY

If angles have to be measured to a higher degree of accuracy than that discussed in previous sections, larger and more expensive theodolites must be employed. All theodolites are basically similar in design and the field procedures used are also much the same.

Fig. 2.82   Observations using a one-second theodolite.

With distance measurement, more variety exists in the possible methods and three will be mentioned.

2.6.1.1 SUBTENSE BAR

A subtense bar has two targets which are accurately fixed at a known distance apart (usually exactly

2 m) at the ends of a bar that can be rotated until it is at right angles to the line to be measured (Fig. 2.83). A third target, in the middle of the bar, is centred over the end of the line to be measured.

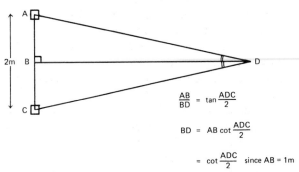

$$\frac{AB}{BD} = \tan\frac{ADC}{2}$$

$$BD = AB\cot\frac{ADC}{2}$$

$$= \cot\frac{ADC}{2} \quad \text{since } AB = 1m$$

Fig. 2.83   Principle of subtense bar measurement.

The measurement taken in the field is the angle at D subtended by the ends A and C of the bar, i.e. the angle ADC. When a 2 m bar is used, the distance is found from the cotangent of half the measured angle and will be the horizontal distance between points B and D, irrespective of the height difference which may exist between the end-points of the line. If a one-second theodolite is used, and if the distance is less than 50 m, an accuracy of about 1 in 5000 can be obtained, but only with repeated, careful measurements of the subtended angle on both F.L. and F.R. For field scientists with little observational experience, this method is not recommended, as small observational errors give very large errors in the result.

2.6.1.2 TELEMETER

The telemeter attachment is set on the front of the objective lens of an ordinary theodolite telescope and a special horizontal staff, held on a tripod, is viewed. Two images of the staff are seen, one directly and the other deflected by the glass wedge in the telemeter attachment through an angle whose tangent is 1/100. The deflected image is displaced with repect to the main image by 1/100th of the distance from the theodolite to the staff. The zero mark on one image points to the reading which must be taken on the other, overlapping image. The slant distance is obtained, and a vertical angle must be measured so that the horizontal distance can be computed.

2.6.1.3 ELECTROMAGNETIC DISTANCE MEASURING (EDM) DEVICES

There are now many electronic systems available for the measurement of distance: some are designed for navigation and have a low accuracy, whereas others are designed for base line measurement and have a very high accuracy; there are both short- and long-range systems; some use visible light, some infra-red radiation and others operate in the microwave region of the electromagnetic spectrum.

The development of EDM has had a great influence on the methods used by the practising surveyor, for instance: trilateration, traversing and radiation, all of which require distance measurement, are now gaining in importance and can be used over much longer distances than were possible with conventional methods. The time spent in the field can be reduced, better and more accurate frameworks can be laid out, and the surveying of specific points by the radiation method can be extended well beyond the range of tacheometry.

The EDM instruments of most interest to the field scientist are those employing radiation in the infra-red part of the spectrum. These instruments are generally compact, relatively cheap, simple to operate and require only an unattended prism as their target. If the lines to be measured are short (less than about 200 m), simple and cheap acrylic reflectors can be substituted for the rather expensive glass prisms provided by the instrument manufacturers.

Infra-red radiation is sent out from the instrument at one end of the line to be measured and is reflected back to the instrument from an inactive reflector placed at the other end of the line. Knowing the speed of this radiation in the atmosphere, it is possible to have displayed the distance between the instrument and reflector. The instrument and reflector must be centred over the endpoints of the line with care, and usually fit on the same tripod or tribrach as the theodolite used to measure the horizontal and vertical angles.

While these EDM instruments are capable of giving very high accuracies, it is their convenience and range of operation which are of more interest to most field scientists. Once set up on a clear line within working range, these instruments output distance within seconds, regardless of the type of terrain or length of line, and are therefore attractive for all but the very shortest of distance-measuring tasks, even if high accuracy is not required.

A great variety exists in the instruments currently available: some give slope distance; some can give horizontal distance automatically; the most sophisticated automatically record the measured data on magnetic tape or solid-state memory units for later computer processing. The cheapest instruments, giving only slant distance and usually mounted on a theodolite, are more likely to be used by field scientists, and these instruments are relatively easy to use if the survey system of which they are a part is well understood. It is easier for an untrained surveyor to obtain observations of distance using a modern EDM instrument than it is for him to obtain good observations of angles using a modern theodolite.

The basic output of slant distance between the instrument and reflector can be converted to a horizontal distance if the slope of the measured line is measured. The horizontal distance D and the difference in height dH between the EDM instrument and its reflector can be computed from

$$D = L \cos V$$
$$dH = L \sin V,$$

where L is the measured slope distance and V is the vertical angle.

Knowing the height of the instrument and reflector above ground and the absolute height of one station, the absolute height of the other station can be determined. This is an application of trigonometric heighting (see sect. 2.3.2), and for long lines the curvature and refraction correction must be applied (sect. 2.3.2.3).

Electronic distance measuring instruments cost between £2000 and £10 000 and are therefore not bought unless a considerable work programme can be set up for them. Several companies now hire these instruments, and this might be attractive to short-term users such as field scientists.

Because modern EDM instruments are very reliable and simple to use, there is a tendency to accept without question any output they give.

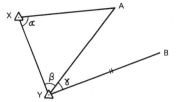

Fig. 2.85   Fixing by intersection and radiation.

The alternative, computational solution is based on the rectangular system of coordinates and a rectangular grid. The latter is drawn up and, using the known coordinates, the points X and Y are plotted. If no coordinates are available for these, X should be assigned an arbitrary value (say 1000-1000) and the direction XY an arbitrary orientation. The position of X can be plotted directly and the coordinates of Y computed by radiation.

Once the grid is drawn and the base points plotted, the coordinates of all other points are then determined by computation, thus allowing all other points to be plotted on the grid. Some methods of obtaining coordinates will now be discussed.

2.6.2.1 RADIATION

If the coordinates of points X and Y are known, the bearing of line XY can be determined (Fig. 2.86).

$$\tan \theta = \frac{\Delta E}{\Delta N}$$

If the bearing of line XY was found to be 140°, say, the reverse bearing will be the bearing of line YX.

$$
\begin{aligned}
\text{Bearing YX} &= \text{Bearing XY} + 180° \\
&= 140° + 180° \\
&= 320°.
\end{aligned}
$$

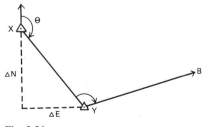

Fig. 2.86

Fig. 2.84   Electromagnetic distance measuring device.

While this faith is justified in most cases, the field scientist must constantly bear in mind that no single observation can ever be fully guaranteed (see sect. 2.1) and that independent checks, repeated measurements and a full understanding of the method of operation are as essential with these instruments as with less sophisticated ones.

2.6.2 INTRODUCTION TO COMPUTATIONAL METHODS OF FIXING POSITION

In Fig. 2.85, point A has been fixed by intersection and point B by radiation. The accuracy of the graphical plotting of both these points will depend on the accuracy with which angles α, β and γ can be set out from the base line XY and on the accuracy of the distance plotting. If the length of the rays XA, YA and YB are large compared to the diameter of the protractor used to plot the angles, these lines will contain errors in their orientation and this is most likely to happen at large scales, where accuracy is of greatest importance.

If the angle at Y between X and B is equal to 120°, the bearing of line YB can be found by adding this angle to the bearing of line YX.

$$\text{Bearing YB} = \text{Bearing YX} + XYB$$
$$= 320° + 120°$$
$$= 440° \text{ (or } 440° - 360°) = 80°.$$

If the distance YB and the bearing YB are known, the difference in easting and northing between Y and B can be calculated as

$$\Delta E_{YB} = D \sin 80°$$
$$\Delta N_{YB} = D \cos 80°.$$

The coordinates of B are then

$$E_B = E_Y + \Delta E_{YB}$$
$$N_B = N_Y + \Delta N_{YB}$$

and point B can then be plotted.

This computation could be repeated for many points surveyed from a single radiation station. Such a repetitive computation is ideally suited to a small programmable pocket calculator (see sect. 2.6.2.7).

2.6.2.2 INTERSECTION

Intersection can be computed from first principles, making use of the sine rule in the intersection triangle, but formulae are available which allow a direct solution to this problem. Only two methods, suitable for use on a small electronic calculator, will be given here.

When the angles for intersection have been measured by theodolite, the coordinates of the new point can be determined in the following manner. With the point P on the left of the line AB (Fig. 2.87), and the coordinates of A and B known

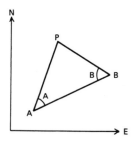

Fig. 2.87

$$E_P = \frac{E_A \cot B + E_B \cot A + N_A - N_B}{\cot A + \cot B}$$
$$N_P = \frac{N_A \cot B + N_B \cot A - E_A + E_B}{\cot A + \cot B}$$

If the bearings from A and B to point P have been computed as a and b, then an alternative solution is given by

$$E_P = \frac{E_A \cot a - E_B \cot b - N_A + N_B}{\cot a - \cot b}$$
$$N_P = N_A + (E_P - E_A) \cot a.$$

2.6.2.3 RESECTION

One method of computing a resection is shown in Fig. 2.88.

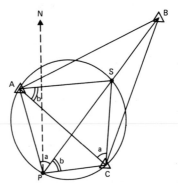

Fig. 2.88

A, B and C are known points, and the position of the unknown point P must be determined using the two angles a and b measured at P. The circle through A, P and C meets PB (or PB produced, if B lies inside the circle) at point S. A and C are joined to S. Angle CAS = angle APS. In triangle ACS, the coordinates of A and C are known, as are the angles a and b. The coordinates of S can therefore be computed by intersection. Knowing the coordinates of S and B, the bearing of line BS can be computed (see sect. 2.6.2.1 and Fig. 2.86), which is also the bearing of line BP.

The bearing of lines PA and PC can then be computed and the coordinates of point P can be found by intersection from any pair of known points (A, B and C).

2.6.2.4 TRILATERATION

If the distances AP and BP from known points A and B to the unknown point P are measured (Fig. 2.88), the coordinates of point P can be calculated by bearing and distance from A or B if the angle BAP or ABP can be found.

With the three sides of any triangle ABC known, the half-angle formula can be used, i.e.

$$\sin A = \frac{2}{bc} \sqrt{s(s-a)(s-b)(s-c)}$$

where A is the angle, a, b and c are the lengths of the sides opposite angles A, B and C and

$$s = \tfrac{1}{2}(a + b + c).$$

Knowing the bearing of line AB (Fig. 2.88) from its coordinates, the bearing of AP or BP can be calculated once angle PAB or PBA is computed. The coordinates of P can then be established by radiation.

2.6.2.5 TRAVERSE COMPUTATION

A traverse can be considered to be a series of radiation computations, each angle and distance being used to determine the difference in eastings and northings between the stations at each end of each line. The misclosure in eastings and northings at the closing station at the end of the traverse must be distributed along the traverse in proportion to the length of each leg. This adjustment can be carried out graphically or numerically.

2.6.2.6 EXAMPLES

Table 2.6 shows the computation and adjustment of the connecting traverse shown in Fig. 2.89.

With traverse points 3 and 4 fixed, other points could then be fixed by radiation, resection, intersection, and trilateration. Table 2.7 shows the computation of the coordinates of point 7 by intersection and Table 2.8 the computation of point 8 by resection.

With the six lines of a braced quadrilateral measured by tape or chain, it is possible to compute the coordinates of three of the points if coordinates for one point and the direction of one line are either known or assumed. This approach might be useful for a small archaeological survey on a local coordinate system, where plotting was to be carried out at a large scale (say 1:50 or 1:100). Table 2.9 gives details of such a trilateration computation.

In all survey calculations, it is wise to construct a diagram as an aid to the calculation. A good diagram, drawn to scale with angles and bearings plotted with a protractor, can help detect gross error in the computations and makes for easier understanding of the computational procedures.

It is necessary to incorporate independent checks into all survey computations, just as much as in survey fieldwork. Calculations should be carried out twice, preferably by different people, and whenever there are two methods available for the computation of a coordinate, each should be used, one as a check on the other.

Fig. 2.89

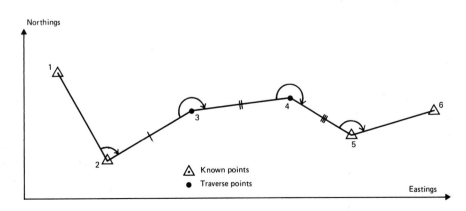

67

Table 2.6

Given data			Measured data		
Point	Eastings	Northings	Angles	2	89° 37′ 10″
2	695.33	296.81		3	200° 21′ 30″
5	2685.22	487.88		4	217° 06′ 00″
				5	132° 44′ 45″
Bearing 2–1	332° 46′ 25″		Distances	2–3	802.22
Bearing 5–6	72° 35′ 30″			3–4	795.36
				4–5	565.09

Computation of bearings

			Corrected angles	Corrected bearings
Bearing	2–1	332° 46′ 25″		
Angle	2	89° 37′ 10″	89° 37′ 05″	
Bearing	2–3	62° 23′ 35″		62° 23′ 30″
Bearing	3–2	242° 23′ 35″		
Angle	3	200° 21′ 30″	200° 21′ 25″	
Bearing	3–4	82° 45′ 05″		82° 44′ 55″
Bearing	4–3	262° 45′ 05″		
Angle	4	217° 06′ 00″	217° 05′ 55″	
Bearing	4–5	119° 51′ 05″		119° 50′ 50″
Bearing	5–4	299° 51′ 05″		
Angle	5	132° 44′ 45″	132° 44′ 40″	
Bearing	5–6	72° 35′ 50″		72° 35′ 30″

Known bearing 5–6   72° 35′ 30″
Angular misclosure   20″
Angular correction 20″/4 = 5″ per angle

Pt	Corrected bearing	Distance	ΔE	ΔN	Prov. E	Prov. N	dE	dN	Eastings	Northings
1	152° 46′ 25″	—	—	—	—	—	—	—	—	—
2	62° 23′ 30″	802.22	710.88	371.77	695.33	296.81	—	—	695.33	296.81
3	82° 44′ 55″	795.36	789.00	100.39	1406.21	668.58	−0.04	+0.05	1406.17	668.63
4	119° 50′ 50″	565.09	490.13	−281.24	2195.21	768.97	−0.09	+0.11	2195.12	768.08
5	72° 35′ 30″	—	—	—	2685.34	487.73	−0.12	+0.15	2685.22	487.88
6										
					dE = −0.12	dN = +0.15				

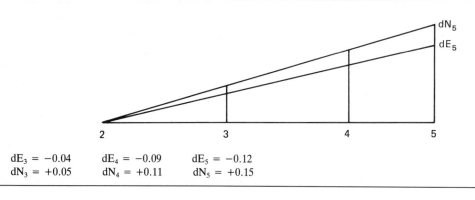

dE₃ = −0.04   dE₄ = −0.09   dE₅ = −0.12
dN₃ = +0.05   dN₄ = +0.11   dN₅ = +0.15

Table 2.7

Given data			Measured data	
Point	Eastings	Northings	Angle A	47° 44′ 20″
3 (A)	1406.17	668.63	Angle B	38° 10′ 15″
4 (B)	2195.12	769.08		
$\Delta E = 788.95$	$\Delta N = 100.45$			

Computation of bearings

$$\tan (\text{bearing } 3\text{–}4) = \frac{\Delta E}{\Delta N} = \frac{788.95}{100.45}$$

Bearing 3–4	= 82° 44′ 39″
Angle at 3 (angle A) =	47° 44′ 20″
∴ Bearing 3–7	= 35° 00′ 19″
Angle at 4 (angle B) =	38° 10′ 15″
∴ Bearing 4–7	= 300° 54′ 54″

Computation of coordinates of point 7

Point	Easting	Northing	Bearing	Cot
A	1406.17	668.63	35° 00′ 19″	1.427 868
B	2195.12	769.08	300° 54′ 54″	−0.598 843

Substitution in bearing intersection formulae gives
$E_7 = 1688.85$
$N_7 = 1072.26$

Check using angle intersection formulae

Point	Eastings	Northing	Angle	Cot
A	1406.17	668.63	47° 44′ 20″	0.908 690
B	2195.12	769.08	38° 10′ 15″	1.272 105

Substitution in angle intersection formulae gives
$E_7 = 1688.85$
$N_7 = 1072.26$

2.6.2.7 THE USE OF ELECTRONIC CALCULATORS AND COMPUTERS

In all computational methods, it is necessary to use trigonometric functions. Reading these functions from tables will make these methods very laborious, especially if many points are to be fixed. The situation can be eased by the use of a modern electronic calculator with function keys. These instruments take much of the toil out of survey calculations and help greatly in reducing the number of errors that are made.

When many computations have to be performed, the use of a programmable calculator is highly recommended. These instruments are not expensive (£15 upwards) and many field scientists now use them for statistical work, and writing short programs for such tasks as converting from polar to rectangular coordinates, reducing tacheometric observations and determining height differences from vertical angles is relatively easy. In fact, many manufacturers provide such programs for interested users.

The use of electronic computers by professional surveyors is now widespread, and although it is unlikely that a field scientist would find it necessary to use a large computer for his survey work, if access to such a machine was readily available, this approach should also be considered. For the methods discussed in this book, the programming is simple and any computing service or agency would be able to advise on suitable programs.

Microcomputers are also widely used in the survey industry and many field scientists have access to this type of computer in their own offices or laboratories. Instruments such as the Commo-

Table 2.8

Given data			Measured Data
Point	Easting	Northing	Angle at 8 between 3 and 4 = 44° 27′ 30″
3	1406.17	668.63	Angle at 8 between 4 and 5 = 54° 07′ 49″
4	2195.12	769.08	
5	2685.22	487.88	

Computation of coordinates of S.

Point	Easting	Northing	Angle	Cot.
3	1406.17	668.63	54° 07′ 49″	0.723 074
5	2685.22	487.88	44° 27′ 30″	1.019 089

Substitution in angle intersection formulae gives
E = 2040.78
N = 1327.78

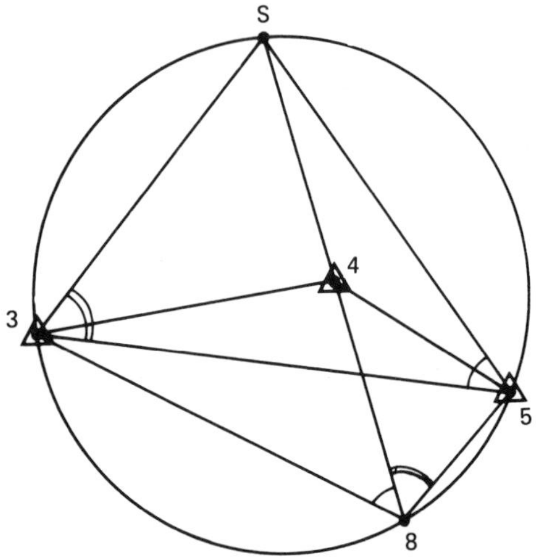

Computation of bearing of S4 (4–8)

Point	Easting	Northing
S	2040.78	1327.78
4	2195.12	769.08

$$\Delta E = 154.34 \qquad \Delta N = -558.70$$

$$\tan (\text{Bearing S–4}) = \frac{\Delta E}{\Delta N} = \frac{154.34}{-558.70} = -0.276\ 248$$

Bearing S–4 = 164° 33′ 26″ = Bearing 4–8

Computation of coordinates of point 8

Bearing 4–8 = 164° 33′ 26″
Bearing 8–4 = 344° 33′ 26″
Angle 483 = 44° 27′ 30″
Bearing 8–3 = 300° 05′ 56″
Bearing 3–8 = 120° 05′ 56″

Point	Easting	Northing	Bearing	Cot.
3	1406.17	668.63	120° 05′ 56″	−0.579 654
4	2195.12	769.08	164° 33′ 26″	−3.619 919

Substitution in bearing intersection formulae gives
E = 2378.58
N = 104.97

dore PET and 64, the Acorn Atom, the Apple and BBC Micro are suitable for all survey computations likely to be carried out by field scientists. Although the programs necessary for survey computations are not complicated, the field scientist should not underestimate the time necessary for the writing and thorough testing of such programs. Unless a great deal of computing work is expected, the use of microcomputers may not give a quicker or easier solution than a calculator if program development has to be carried out by the field scientist himself. The use of program packages, available from some manufacturers and software houses, could be considered as an alternative to writing programs. Again, it is usual for some difficulties and delays to occur before such packages can be used successfully.

The use of electronic computers can bring enormous benefit to the user, but these advantages do not come automatically. The field scientist must set aside time to study and understand the workings of the computer and the programs if he is to exploit the full potential of these systems. Moreover, the use of the computer can never compensate for poor

Table 2.9

Measured data	

AB = 44.85 m
BC = 24.93 m
CD = 28.46 m
DA = 28.31 m
AC = 45.39 m
BD = 43.67 m

Assume that coordinates
of point A are

100 m E 100 m N

Assume that bearing AB = 90°.

In triangle ABC,
a = 24.93, b = 45.39, c = 44.85
$s = \frac{1}{2}(a + b + c) = 57.58$

$\sin A = \dfrac{2}{bc} \sqrt{s(s - a)(s - b)(s - c)}$

        = 0.530 976
A   = 32° 04' 17"
Bearing AC = 57° 55' 43"
E_c = E_A + AC sin 57° 55' 43"
    = 100 + 38.46
    = 138.46
N_c = N_A + AC cos 57° 55' 43"
    = 100 + 24.10
    = 124.10

In triangle ABD,
a = 43.67, b = 28.31, d = 44.85
$s = \frac{1}{2}(a + b + d) = 58.415$

$\sin A = \dfrac{2}{bd} \sqrt{s(s - a)(s - b)(s - d)}$

        = 0.934 203
A   = 69° 05' 59"
Bearing AD = 20° 54' 01"
E_D = E_A + AD sin 20° 54' 01"
    = 100 + 10.10
    = 110.10
N_D = N_A + AD cos 20° 54' 01"
    = 100 + 26.45
    = 126.45

Check – Distance DC from coordinates
$\sqrt{*138.46 - 110.10)^2 + (124.10 - 126.45)^2} = 28.46$
∴ Distance DC from computed coordinates is equal to measured distance DC, not yet used in the computation.

fieldwork and bad observations: great care must be taken to ensure that the quality of the input is suitable for the processing to be carried out.

2.7 POSTSCRIPT

Because of the infinite variations found in the real world, few field situations encountered by the field scientist will be covered exactly by a textbook description of method and technique. No two survey jobs are quite the same, much to the relief of most professional surveyors for therein lies much of the interest. However, the methods and instruments described in this section are capable of providing solutions to many survey problems other than those specifically mentioned. Also, instru-

ments and methods can be used in combinations different from those described in detail in the text. With a little experience, the field scientist will be able to select from the wide range of possible survey methods the one which will best suit his needs in a particular situation. In this context, the readers may view the chapters of this section as describing model solutions which he will adapt to his own requirements.

Such modifications, along with purely pragmatic solutions, are common in surveying but it is important that such actions be carried out within the framework of the principles and practices described in this section.

Effective surveying for the field scientist is a combination of common sense, especially with regard to logistics, experience and a sound grasp of principles, allied to a knowledge of the more common methods.

FURTHER READING

Bannister, A. and Raymond, S., 1984, *Surveying* (5th edn). Pitman.

Burnside, C. D., 1982, *Electromagnetic Distance Measurement* (2nd edn). Granada.

Ingham, A. E., 1984, *Hydrography for the Surveyor and Engineer* (2nd rev. edn). Crosby, Lockwood and Staples.

Oliver, J. G. and Clendinning, J., 1978, *Principles of Surveying* (4th edn, vol. 1). Van Nostrand-Rheinhold.

Chapter 3 AERIAL SURVEYING TECHNIQUES

3.1 IMAGE FORMATION AND EVALUATION

Aerial photography is one of a group of techniques, known collectively as 'remote sensing', in which information about the environment is recorded from a distance, usually from sensors carried in an aircraft or spacecraft. An aerial photograph is a record at an instant in time of the light-reflecting properties of such features of the terrain as are detected by the sensor (the film emulsion). The nature of the image depends largely on the properties of the film, the reflectance characteristics of terrain features and the condition of the atmosphere at the moment of exposure. The amount of detail recorded varies with the scale of the photography and the resolving power of the camera–film system.

Since the landscape is the subject of multi-disciplinary studies, it is not surprising that a photographic analogue of the landscape attracts a variety of users. It is the inter-disciplinary utility of the aerial photograph that has led to its widespread adoption in all types of landscape studies.

The utilisation of aerial photography can be divided between two main activities that in practice are complementary. 'Photo-interpretation' is a qualitative aspect concerned with the identification of features and an assessment of their significance, and 'photogrammetry' is a quantitative aspect concerned with the accurate measurement of features recorded by photography, though not necessarily aerial photography.

'Image interpretation' is a more general term which is used to indicate that the sensor employed need not be a camera–film system but might, for example, be a thermal infra-red sensor or a micro-wave sensor producing a 'photo-like' image as one form of data display. These other forms of remote sensing are considered in Chapter 4.

3.2 PHOTO-INTERPRETATION

As defined by the American Society of Photogrammetry, photo-interpretation is 'The act of examining photographic images for the purpose of identifying objects and judging their significance'. The success in carrying out this task will depend in large measure on the prior training and experience of the interpreter in the discipline(s) relevant to the problem under consideration. Notwithstanding the obvious expectation that, for example, someone with the relevant training in forestry will make a more accurate interpretation of a forested area than someone without such training, there are some aspects of photographic characteristics which can assist in the adoption of a systematic approach to interpretation.

Most textbooks dealing with this subject come up with a similar listing of the factors which are most relevant in photo-interpretation, first promulgated in the *Manual of Photographic Interpretation*: shape, size, shadow, tone, texture, pattern, site, situation and association. (Colwell (ed.) 1960; Dickinson 1979; Paine 1981; Avery and Berlin 1985.) Although each of these factors may have an influence in the correct identification and classification of an object, their relative importance will vary from feature to feature. The correct identification is likely to be the result of a consideration by the interpreter, albeit subconsciously, of a number of different factors. Identification by 'convergence of evidence' is the usual description of this approach.

Shape
The configuration or outline of an object may help in its identification. However, the interpreter may be unaccustomed to the vertical, or plan view, a dilemma which may be resolved if the photograph scale is known.

Size

The physical dimensions of a feature may be a distinctive attribute, and may assist where the shape is inconclusive. This requires knowledge of the nominal photograph scale so that photograph measurements can be converted to the real ground size of a feature. This apparently simple factor may be of crucial significance when, for example, trying to decide on the likely identity of animals, without the aid of much ancillary information.

Shadow

Where illumination conditions result in shadows on the photograph, the side-on or elevation view which this provides may be more helpful in defining some objects than their vertical or plan view (for example, bridges, towers or chimneys).

Tone (or colour)

In a fundamental sense, features on a photograph are detectable only because differences in tone or colour result from the differing spectral reflectance properties of the features in the scene. A basic general knowledge of the reflectance properties of different ground targets (both natural and man-made) is an aid to interpretation. Additionally, specialist knowledge of how tones are likely to alter, for example, as a result of vegetative growth, or the effects of drought, disease or ageing on plants, enables the user to make a more meaningful interpretation of tonal variations. Comparison of tones should generally be made only within a photograph or set of adjacent photographs which have been subjected to the same processing conditions, since variations in processing can affect the absolute tonal rendition of features on photographs. Although seldom available, the ideal is to have a set of dark and light targets of known ground reflectances included in each film sortie in order to permit calibration of the resulting image tones.

Texture

The rate of change of tone within the image is known as texture. This can assist, in particular, in the identification of vegetation cover types, but some account must be taken of the photograph scale. For example, at large scale a wooded area may appear to have a coarse texture, but from a very high altitude the resulting small-scale photography may depict the same area in a smooth texture.

Pattern

Pattern refers to the spatial arrangement of a feature, and may be a distinctive identifying characteristic. For example, the planting pattern of many crops is distinctive and in the case of fruit trees this may be a clue to definitive identification. Regular geometric patterns are usually associated with man's impact on the landscape, although some natural patterns may act as surrogates for an underlying characteristic, such as the relationship which sometimes exists between drainage pattern and rock types.

Site, situation and association

The site is the piece of terrain occupied by the feature of interest. In the case of many man-made features, the site was frequently carefully selected, and especially for many historical or archaeological features a knowledge of such favoured sites can be of crucial importance in helping to identify a feature. Similar reasoning can be applied, for example, to features associated with economic activity or in the detection and recognition of certain physical features, such as sandbanks along a meandering river course. The three-dimensional image provided by stereo-viewing is particularly important in site recognition.

The situation of a feature refers to the wider setting within the photograph, or indeed within the region depicted on the photograph. Knowledge of the geographic location of the photographed area acts as a filter on the interpreter's fund of knowledge, screening out a large number of unlikely possibilities and concentrating attention on a smaller number of probabilities. Association is the word used to describe the line of reasoning whereby a feature may not be conclusively identified by itself, but by encompassing a group of features associated with a particular function a more certain identification is often possible.

This may be the case, for example, when concluding that the association of a riverside site, railway sidings, canal basin, coal heaps and a large building with smoke stacks suggests that there is a long-established heavy industrial plant, which was initially dependent on water for the processing of imported raw material, and coal which was initially brought in by canal and which is now transported by rail. The final deduction and classification of a feature should be as a result of the systematic use of several of the above factors. The convergence of evidence provided by a multi-factor approach usually gives the most reliable interpretation. This is especially so with small-scale photography where the instant feature recognition, which is so often applied to large-scale photography, is not so readily used.

3.3 AERIAL CAMERAS

It is possible to carry out a certain amount of qualitative interpretation of any kind of aerial photography, regardless of the type of camera used. Special requirements should be satisfied, however, in the taking of the photographs if they are to be used for measuring purposes, especially if a map is to be the end product.

Just as the user of maps in an atlas must be aware of the properties of projections if he is to avoid drawing incorrect conclusions about, for example, shapes or areas on the map, any user of aerial photography must be familiar with certain properties of aerial photographic imagery if he is to make proper use of the medium. The type of film emulsion – panchromatic, infra-red, colour or false colour – determines to a great extent the nature of the qualitative information recorded on an aerial photograph.

The optical system of the camera can adversely influence the image position on the photograph by deflecting an incident light ray from the geometrically correct path. This 'distortion', which is characteristic of the camera lens, is most serious with cameras used for 'reconnaissance' photography, but is almost negligible in aerial 'survey' cameras. The usefulness of the photographs for measurement purposes will therefore depend on whether a reconnaissance camera or an aerial survey camera was used to take them. Reconnaissance photography may be taken to include all aerial photography not done specifically for measuring purposes. Reconnaissance cameras may carry out continuous strip photography, have focal-plane shutters, or have lenses of very long focal length and high distortion (Figs. 3.1 and 3.2 and 3.3), all of which make the resulting photographs unsuitable for all but the most approximate measurements.

3.3.1 RECONNAISSANCE CAMERA

3.3.1.1 HIGH ALTITUDE

A camera with a very long focal length is used ($F = 50$ cm, or even $F = 150$ cm) in order to obtain the largest possible photograph scale. The lens must be able to separate the images of small ground features, i.e. a high resolving power is required. The metric accuracy of the image is unimportant, so that large radial lens distortions are usually present. A focal-plane shutter is used, primarily for its reliability and high shutter speeds.

Fig. 3.1 Focal length.   Parallel light rays, from infinity, passing through a thin lens, will come to a point in the focal plane.

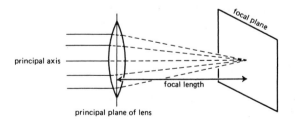

The FOCAL LENGTH of the lens is the length of the perpendicular from the focal plane, along the principal axis of the lens to the lens centre.
In aerial photography the object is at almost infinite distance.
The aerial camera is therefore constructed to focus on an object at infinity, by making the perpendicular distance from the film plane to the lens centre equal to the focal length of the camera lens.

Fig. 3.2 Characteristics of some aerial cameras.

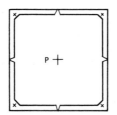

FORMAT (picture size) most common is 23 x 23 cm²
Fiducial marks, crosses in corners or thorns on centre of
sides, locate the point (P) (see Fig 3.17).
In some parts of Europe a common format is 18 x 18 cm²,
for which the corresponding focal lengths are 210mm for N A
and 115mm for W A   (see below)

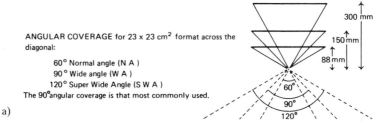

ANGULAR COVERAGE for 23 x 23 cm² format across the
diagonal:

 60° Normal angle (N A)
 90° Wide angle (W A)
 120° Super Wide Angle (S W A)
The 90° angular coverage is that most commonly used.

a)

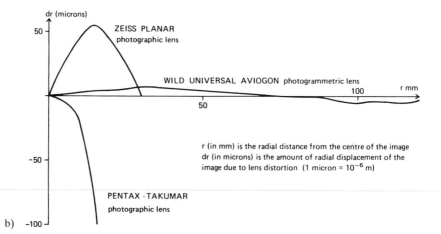

b)

r (in mm) is the radial distance from the centre of the image
dr (in microns) is the amount of radial displacement of the
image due to lens distortion (1 micron = 10^{-6} m)

3.3.1.2 LOW ALTITUDE AND VERY HIGH FLYING SPEED

In order to avoid blurring of the image, image movement compensation (IMC) is achieved by moving the film across the focal plane in step with the aircraft's speed over the terrain. The terrain is photographed continuously on the moving film and there is thus no single centre of perspective.

3.3.2 METRIC CAMERA

Photography for topographic mapping should be done with an aerial survey camera, which ensures a high geometric accuracy in the image. Such a camera has the following features.

1. A lens with a low radial distortion (less than 10 μ).
2. Fiducial marks to locate the principal point (see Figs. 3.2 and 3.17).
3. Focal length of lens calibrated and known correct to 0.01 mm (see Fig. 3.2).
4. A film-flattening device to ensure a flat film surface at the moment of exposure.

The most common photogrammetric use of aerial photographs is in the production of medium- and small-scale topographic maps. Photography for this purpose is done with a specially designed aerial survey camera equipped to minimise such errors as, for example, are due to lens distortion. For map production the photographs are used in plotting

Fig. 3.3 Reconnaissance and metric cameras.

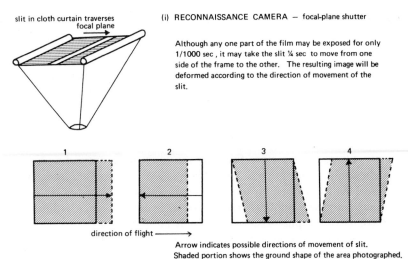

slit in cloth curtain traverses focal plane

(i) RECONNAISSANCE CAMERA — focal-plane shutter

Although any one part of the film may be exposed for only 1/1000 sec , it may take the slit ¼ sec to move from one side of the frame to the other. The resulting image will be deformed according to the direction of movement of the slit.

direction of flight ⟶

Arrow indicates possible directions of movement of slit.
Shaded portion shows the ground shape of the area photographed.
Since there is no instantaneous exposure of the photographed area there is no single perspective centre.

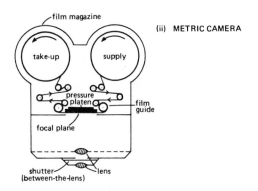

(ii) METRIC CAMERA

film magazine

take-up   supply

pressure platen

film guide

focal plane

shutter (between-the-lens)   lens

machines, which can accommodate only certain formats (size of photograph) and focal lengths (Fig. 3.2). Since the widespread adoption in recent years of photogrammetry as the basic technique for the production of topographic maps, a great number of high-quality photographs that were originally taken for mapping purposes have become available to the field scientist for photo-interpretation or for measurement purposes.

A distinction is made between types of aerial photograph on the basis of the attitude of the camera axis at exposure. If the axis is intentionally tilted by, say, 20° from the vertical, the photograph is termed an 'oblique' ('high' oblique if the horizon is recorded and 'low' oblique if it is not). Almost without exception, aerial photography for surveying purposes is done with the camera axis vertical, or as near vertical as is possible. In practice, it usually comes within 1–2° of vertical. It is possible to use oblique photography for measurements and mapping, but there are considerable difficulties owing, for example, to the continuously changing scale from foreground to background, and special procedures must therefore be adopted. Although the oblique photograph may give a conventionally more familiar view than the vertical photograph, and be useful for illustrative purposes, it is of much less value for detailed interpretation. For these reasons the techniques to be described are mainly for use with near-vertical photography (see Fig. 3.14).

3.3.3. BINOCULAR VISION AND STEREOSCOPY

Man's ability to see in three dimensions is a result of having two eyes that record slightly different images of the objects around him. These two

Fig. 3.4 Binocular vision.

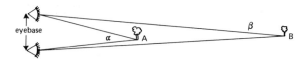

CONVERGENT ANGLE α is greater than β therefore A is closer than B.

With good eyesight it is possible to detect differences between α and β at about 30 secs of arc.
Beyond about 200 feet the impression at depth falls away and the scene becomes 'flatter'.

slightly different images are fused by the brain to give the impression of depth (Fig. 3.4).

An essential feature of aerial photography for mapping purposes, although not always a feature of reconnaissance photography, is that it must be possible to view an optical model of the terrain in three dimensions. This is made possible if geometrical conditions similar to natural binocular vision, described above, are satisfied in the taking of the aerial photographs. At successive exposures the aerial camera positions are analagous to the separation of the eyes in normal vision (Fig. 3.5). If the photographic sequence is arranged so that a portion of the terrain between the two camera stations is recorded on successive exposures, two slightly different views of the same area of terrain, called the 'overlap', are recorded on the photographic emulsion. If the pair of photographs are viewed under suitable conditions, an optical model of the area of overlap will be seen in relief. The condition to be satisfied is for each eye to have a slightly different view of the same area of terrain; this can be achieved if the left eye views only the left photograph and the right eye views only the right photograph of the overlap (Fig. 3.6).

A three-dimensional, or stereoscopic, image of the terrain cannot be created from aerial photographs unless the conditions for binocular vision are reproduced in their taking. The dimension of height not only enhances the interpretation of many features but also allows the photogrammetrist the possibility of measuring heights within the stereoscopic model, using only a minimal amount of ground-derived information (normally four known height points for one overlap).

Fig. 3.5 Aerial photography – conditions for stereoscopy.

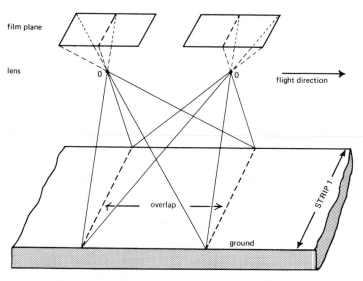

Minimum overlap for stereoscopic cover is 50%. In practice, to avoid gaps, a 60% overlap along the flight direction and a 30% overlap between adjacent strips is applied.

Fig. 3.6 Giant eye-base.

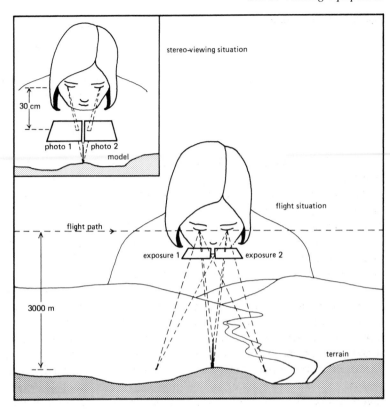

3.4 STEREO-VIEWING EQUIPMENT

3.4.1 THE LENS STEREOSCOPE

The simplest, and least expensive, piece of equipment for viewing the three-dimensional image from overlapping photographs is the lens stereoscope (Fig. 3.7). It consists of a pair of lenses mounted on a frame, which is supported over the photographs by thin metal legs.

The lens separation may be fixed for the average human eye separation (eye-base), but more usually the separation may be varied to suit the eye-base of the user. When the lens stereoscope is set up, the distance from the lens to the photograph is equal to the focal length of the lenses, so that the photographic image will appear sharp in the focal plane of the lenses. The image appears enlarged, usually 2× to 2.5×. The user has to arrange the photographs below the lenses so that the appropriate eye is viewing the same detail in the appropriate photograph at the same time. It is necessary

to place one photograph on top of the other so that the separation of corresponding images is about the same as the lens separation. As a consequence, only a limited portion of the photograph can be viewed at any time, and the photographs have to be readjusted often if the whole overlap area is to

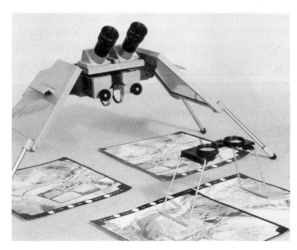

Fig. 3.7 Lens stereoscope and mirror stereoscope.

be viewed stereoscopically. With 23×23 cm² aerial photographs there is a narrow area in the centre of the overlap that cannot be viewed in stereo unless one of the prints is bent to reveal the corresponding images in that area. After considerable practice many users develop the ability to view overlapping photographs in stereo unaided, by simply holding the photo-pair in front of the eyes, with the photographs arranged so that corresponding details are at a separation equivalent to the eye-base.

The portability of the lens stereoscope makes it suitable for use in the field. Its main limitations are the inconvenience of rearranging the photographs, the difficulty of annotating them, the relatively low magnification and the tendency with the inexperienced user for each eye to see both photographs.

Fig. 3.8 Mirror stereoscope with parallel guidance (Hilger and Watts, Type SB 180).

3.4.2 THE MIRROR STEREOSCOPE

Most of the limitations of the lens stereoscope are overcome in the mirror stereoscope. This stereoscope consists of two pairs of parallel mirrors (one large and one small mirror in each pair), arranged so that the outer (large) mirrors are inclined at 45° to the horizontal, one over each photograph. The image on each photograph is reflected from the large mirror, then from the small mirror, to appear as two separate images that are viewed through lenses placed vertically over the small mirrors. The small viewing lenses can be replaced by a binocular attachment, to permit magnification of the images, commonly to 4× or 8×. Apart from the increased magnification, the mirror stereoscope has the advantage that the optical paths from each photograph are physically separated. Once correctly set up for viewing, the photographs can be taped to the table, and viewing of different portions of the overlap is achieved by lifting the mirror stereoscope and readjusting it over the required section of the overlap. In one design of mirror stereoscope

(Fig. 3.8) the scanning of the overlap is achieved by moving the pair of photographs together, on a moveable base, below the large mirrors. The correct relative setting of the photographs is maintained by a parallel guidance, to which the base is attached. Small relative movements of the photographs can be made at each new location in order to attain good stereoscopic fusion of the images.

3.5 PHOTOGRAPHIC EMULSIONS

The wavelengths of visible light occupy a restricted portion of the total spectrum of electromagnetic energy in the atmosphere (Fig. 3.9).

Although a photographic emulsion can sense differences in the reflectance of light from the earth's surface, all emulsions are not sensitive to the same range of wavelengths of light, and indeed some emulsions (e.g. infrared) are sensitive to part of the non-visible portion of the electromagnetic spectrum.

For the purposes of image evaluation the films used in aerial photography can be divided into two

Fig. 3.9 The electromagnetic spectrum.

wavelength							
3×10^{-6} 3×10^{-5}		1×10^{-2}	0.4 0.7	1.5 microns	1 mm	0.8 m	3×10^{6} m
γ rays	x rays	U V	VISIBLE / near	INFRA-RED medium and far	microwave	v h f to l f	

v h f = very high frequency
l f = low frequency

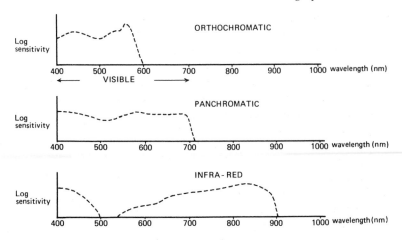

Fig. 3.10 Sensitivity curves of film emulsions.

categories: monochrome films, in which the image is recorded in varying shades of grey (e.g. orthochromatic, panchromatic and infra-red films); and colour films, in which the images are differentiated on the basis of colour differences on the photograph (e.g. true-colour and false-colour films). Within each category further differences exist between films, mainly because of variations in the spectral sensitivity of the emulsions (Fig. 3.10). A consequence of the differences in spectral sensitivity of emulsions is that orthochromatic, panchromatic and infra-red films will display certain unusual image characteristics, which the user should appreciate if he is to gain the fullest value from the photographs.

Orthochromatic film
This type of film occupied an important role in the historical development of black and white emulsions. It has a peak sensitivity in the green portion of the spectrum (Fig. 3.10) and has potential for emphasising green detail, e.g. vegetation cover. It is also the best monochrome emulsion for penetrating water to reveal bottom detail. Despite these attributes, however, it is seldom used today because of its limited sensitivity range.

Panchromatic film
This is sensitive to the full range of visible light, but lacks the marked peak in sensitivity to green displayed by orthochromatic film. It is the film in most common use for aerial photography, owing to its extended sensitivity over the visible spectrum. A yellow (minus blue) filter is almost invariably used with aerial panchromatic film. The function of

a filter is to cut out the wavelengths of light that are shorter than the colour of the filter. For instance, the yellow filter cuts out most of the short-wavelength blue light, which is scattered in the lower layers of the atmosphere and produces a haze effect.

Infra-red monochrome film
This has a broad sensitivity in the visible region (see Fig. 3.10) and is also sensitive to reflected energy beyond the visible, in the near infra-red portion of the spectrum. Owing to the latter most useful quality, this film is commonly used with a deep red filter, which keeps visible light from it. Since the eye is not sensitive to infra-red wavelengths, an infra-red photograph displays characteristics that are not visible on a panchromatic photograph. It is found, for example, that many types of vegetation have different reflectance characteristics in the infra-red and the visible parts of the spectrum. The range of reflectance of vegetation is greatly extended in the infra-red, making differentiation of tree species, for example, much easier than with panchromatic film (Fig. 3.11). Vegetation generally has a higher infra-red reflectance than non-vegetated areas, as have deciduous trees compared with conifers, so that a light–dark tonal contrast enables the user readily to identify this basic division of woodland. Water surfaces absorb most of the incident infra-red radiation, and so usually appear black on infra-red photographs. This is a useful feature in areas of delta, swamp or saltmarsh for the clear delineation of water bodies or stream courses.

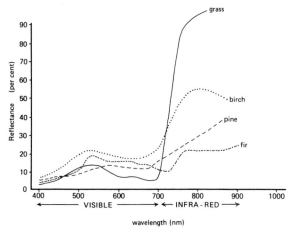

Fig. 3.11 Spectral reflectance signatures of some vegetation types.

On colour films, the emulsion is composed of several layers (usually three) sensitive to primary colours. Intermediate colours are formed by chemical couplers reacting with the three main layers. *True-colour film* consists of blue, green and red sensitive layers. In *false-colour film* the blue sensitive layer is replaced by an infra-red sensitive layer. The green-sensitive and red-sensitive layers produce yellow and magenta colours where unexposed, as a consequence of a colour shift in the sensitive layers (compare Figs 3.12 and 3.13).

In the processing of false-colour film, images do not appear in their natural colours, for example: green and red layers on processing appear respectively blue and green in the final print or transparency. Many of the aforementioned characteristics

Blue sensitive	Yellow forming Layer
Filter	Filter
Green sensitive	Magenta forming layer
Red sensitive	Cyan forming layer
Base	

TRUE COLOUR
Yellow filter to prevent blue light affecting the green and red sensitive layers. Each layer is developed out in three complementary colours (the subtractive primaries). Blue light will remove the image from the blue sensitive layer leaving magenta and cyan in the other two layers. On viewing, magenta and cyan combine to form blue.

Fig. 3.12 Sensitised layers of true-colour film.

Infra-red sensitive	Cyan forming layer
Green sensitive	Yellow forming layer
Red sensitive	Magenta forming layer
Base	

FALSE COLOUR
Blue sensitive layer replaced by Infra-red sensitive layer which gives a cyan positive image on development. If I R radiation affects part of the layer, the image is removed and, as with true colour, the final colour, seen on viewing, is formed by the combination of the remaining two layers, ie I R appears yellow and magenta (red colour). Green and red sensitive layers will appear as blue and green respectively in the final image, hence the name false colour.

Fig. 3.13 Sensitised layers of false-colour film.

of monochrome infra-red film are displayed by the false-colour film and, in addition, the range of colours is much wider than the range of grey tones on the monochrome film.

While colour photography is costlier than monochrome, it generally gives much more information. With true-colour film the greater speed of interpretation is also an advantage. Colour films are not widely used for mapping, although the US National Ocean Survey (formerly the Coast and Geodetic Survey) uses true-colour film for the mapping of coastal areas, and in Great Britain the Ordnance Survey has carried out tests on the metric accuracy of true-colour film.

The user of aerial photography should be aware of the intricacies of the medium with which he is dealing. Furthermore, since an interpretation of features is required before they can be mapped, a knowledge of the unusual features of different emulsions is a prerequisite for the optimum use of aerial photography for mapping.

3.6 PLANIMETRIC MAPPING BY AERIAL PHOTOGRAPHY

The methods of establishing plan position and of preparing a planimetric map by ground surveying techniques are described in Chapter 2. As we have said in Chapter 1, it is also possible to prepare a planimetric map of a ground area from aerial photographs – in fact by photogrammetry. As with ground surveying techniques, however, it is necessary to establish a framework of ground control points before detail mapping can proceed. The photogrammetric method is therefore not totally divorced from ground surveying measurement, since the control points will usually be established and determined in the field. If a pre-existing map or plan is available at a suitable scale, it might be used as a control base.

3.6.1 BASIC REQUIREMENTS FOR PHOTOGRAMMETRIC MAPPING

While it is possible, given some control points, to derive measurements from almost any type of photography, the number of sources of error will be minimised, and the task of transforming the positions of photo-images into planimetric map positions greatly simplified, if the following conditions are observed.

1. The photography is done with a metric camera (see Fig. 3.3).
2. The camera axis is kept vertical (Fig. 3.14).
3. Successive photographs overlap by about 60 per cent along the flight direction (Fig. 3.15).

Although a near-vertical aerial photograph may superficially resemble a planimetric map, there may be considerable plan errors in the photo-images, through failure to satisfy the ideal conditions mentioned above, and also because of inherent differences between the geometric construction of an aerial photograph and a map.

Put in its simplest form, a map is an orthogonal projection of a portion of the earth's surface on to a horizontal datum, whereas an aerial photograph is a central (or perspective) projection (Fig 3.16). Since the aerial photograph is a central projection, the photo-images of ground features suffer displacements whenever there is any ground relief in the photographed area and whenever aircraft movement causes the camera axis to be tilted away from a truly vertical position. The photo-images suffer no displacements from these effects only in the rare situation when perfect vertical-axis photography is obtained of flat ground. In practice, all aerial photography contains some tilt- and relief-displacement errors. The preparation of a plani-

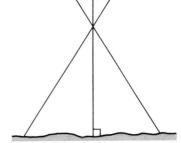

True vertical: camera axis perpendicular to ground surface

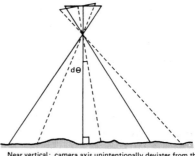

Near vertical: camera axis unintentionally deviates from the true vertical by a few degrees (dΘ)

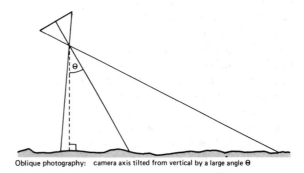

Oblique photography: camera axis tilted from vertical by a large angle Θ

Fig. 3.14 Types of aerial photograph defined by the attitude of the camera axis.

metric map from nominally 'vertical' aerial photographs must therefore include corrections for these image-displacement errors.

3.6.2 CAMERA GEOMETRY AND IMAGE DISPLACEMENTS

The types of camera that may be used for taking aerial photographs are mentioned in Section 3.3. If a metric aerial survey camera is being used, the radial lens distortion effects may be neglected

Fig. 3.15 Overlap requirements for complete photographic coverage.

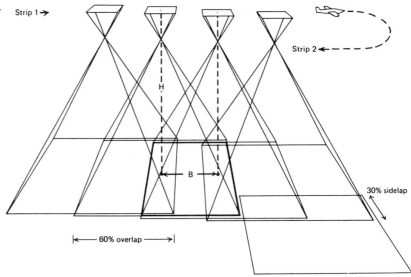

Fig. 3.16 Fundamentals of the geometric construction of a map and an aerial photograph.

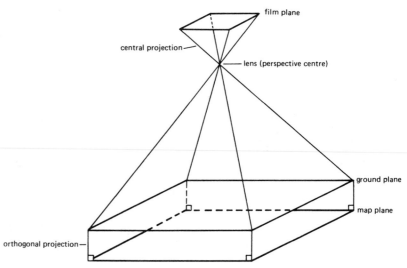

(Fig. 3.2). The nature and cause of the image displacements on the aerial photograph can then best be understood by considering the geometry of the relation between camera and ground at the instant of photography. In a consideration of image displacements on an aerial photograph there are three important photo-points: the principal point; the nadir point; and the isocentre (Fig. 3.17). An appreciation of the definitions of these three points is basic to the understanding and correction of the image-displacement errors.

3.6.2.1 RELIEF DISPLACEMENTS

As can be seen from Fig. 3.18, a vertical feature on the ground AA′ will appear as a displaced linear image on the aerial photograph aa′. More precisely, any point on the ground that has a location and elevation different from the ground plumb-point C will show an image displacement on the photograph.

From Fig 3.18, by similar triangles, it can be seen that

Fig. 3.17 Principal point (p), nadir point (n) and isocentre (i).

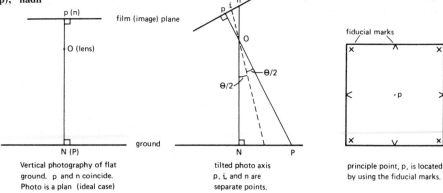

Vertical photography of flat ground. p and n coincide. Photo is a plan (ideal case)

tilted photo axis p, i, and n are separate points.

principle point, p, is located by using the fiducial marks.

The principal point (p) is the point on the image plane at the foot of the perpendicular passing through the centre of the lens.

The nadir point (n) is the point where the vertical line, from the ground, through the lens centre meets the image plane.

The isocentre (i), lying on the line pn, is where the bisector of angle p O n meets the image plane.

$$\frac{bb'}{cb'} = \frac{BB''}{OC} = \frac{BB'}{(H - h)}$$

Now, $bb' = de$ (the image displacement), and $cb' = r$ (radial distance to the top of the image from the photograph centre).

Therefore, $\dfrac{de}{r} = \dfrac{\Delta h}{(H - h)}$

which may be written as

$$de = r \frac{\Delta h}{(H - h)} \tag{1}$$

$$\text{or } \Delta h = (H - h) \frac{de}{r} \tag{2}$$

From formula (1) it can be seen that the image displacement due to relief is directly proportional to the radial distance r of the top of the image from the centre of the photograph, and the ratio of the height of the feature above ground to the ground clearance of the aircraft $\dfrac{\Delta h}{(H - h)}$. From formula (1) the magnitude of image displacement may also be computed.

Table 3.1 shows the values of de for various ratios of ground relief (Δh) to flying height over ground (H − h), and for two radial distances r – near the edge of a photograph on the centre of a side (11 cm), and in the extreme corner of a

Table 3.1

$r \Big/ \dfrac{\Delta h}{(H - h)}$	5 per cent	10 per cent	15 per cent	20 per cent
11 cm	0.55 cm	1.10 cm	1.65 cm	2.20 cm
15 cm	0.75 cm	1.50 cm	2.25 cm	3.00 cm

photograph (15 cm) – assuming a standard aerial photograph of side 23 cm.

Table 3.1 demonstrates that for photographs taken where the ground relief is 20 per cent of the ground clearance of the aircraft, then, if the maximum relief is imaged in the corner of the photograph, the image will be displaced from its correct plan position by 3 cm. For this reason, it is not possible to trace features directly from an aerial photograph and present the result as a map. Apart from being incorrect, boundaries traced from such vertical aerial photographs will be found not to register with their continuations on adjacent photographs. In order to achieve good registration of continued lines and boundaries, it is necessary to compensate for the effects of the relief displacements.

On aerial photography of urban areas, where there is ground relief of buildings, the effect is very noticeable (Fig. 3.19). On aerial photography of upland areas, however, there are seldom any

Fig. 3.18 Linear image displacements on a photograph due to relief.

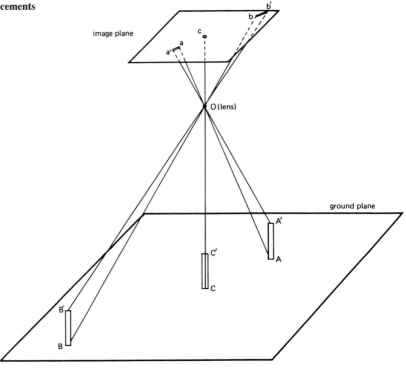

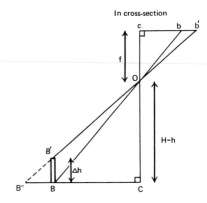

Δh = object height above ground
H = aircraft height above sea level
h = ground height at C above sea level

simple vertical structures to draw attention to this effect. Nevertheless, relief displacements occur at any point that is higher or lower than the ground point vertically below the aircraft.

The point on the photograph at which radial relief displacements originate is the *nadir* point. From a consideration of Fig. 3.18 it can be seen that any object lying vertically under the camera lens will suffer no relief displacement.

3.6.2.2 TILT DISPLACEMENTS

Where the camera axis deviates from the vertical, the photo-images will suffer displacement. It can be shown (Schwidefsky, 1961) that the tilt displacement of a point is radial from the isocentre, and the magnitude dt of the displacement is given by the formula

$$dt = \frac{r^2 \sin \theta}{f},$$

Fig. 3.19 Object height from the radial displacement of the image.

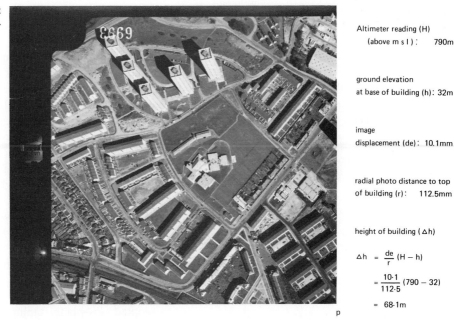

Altimeter reading (H)

  (above m s l) :   790m

ground elevation
at base of building (h): 32m

image
displacement (de): 10.1mm

radial photo distance to top
of building (r) :   112.5mm

height of building (∆h)

$$\Delta h = \frac{de}{r}(H-h)$$

$$= \frac{10 \cdot 1}{112 \cdot 5}(790-32)$$

$$= 68 \cdot 1 m$$

p

where: r = radial distance on the photograph from
the isocentre;

θ = angle of tilt; and

f = focal length of lens.

The most disturbing consequences of tilt displacements are the scale changes that are produced across the photograph. These effects may be appreciated most clearly by considering the appearance on a tilted photograph of some regular feature such as a grid. For convenience, the tilt may be considered to be acting only along the direction of flight, and then only at right angles to the flight direction (Fig. 3.20). In practice, tilt will tend to be of unknown magnitude and will act in a random direction.

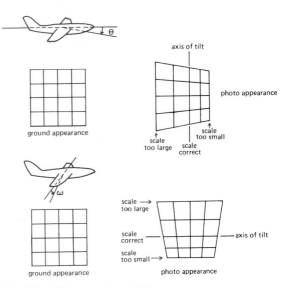

Fig. 3.20 Effects of aircraft tilt on photo-image scale.

3.6.2.3 RADIAL LINE ASSUMPTION

In producing a planimetric map from aerial photographs one must correct the image displacements due to relief and tilt. The points of zero displacement are the nadir point (for relief displacements) and the isocentre (for tilt displacements) so that, strictly speaking, any methods for the removal of relief- and tilt-displacement effects should be referred to the nadir point and the isocentre as origins. It is normally very difficult to locate either the nadir point or the isocentre on an aerial photograph, so that in practice the 'principal point', which is readily located by the fiducial marks, is assumed to be the origin of these radial displacements. This is termed the 'radial line assumption', and is reasonably correct for tilts less than 3° and for ground relief less than 10 per cent of flying height.

3.6.2.4 CORRECTING ERRORS DUE TO RELIEF DISPLACEMENT

When photographed from camera stations O_1 and O_2, a vertical object AA' on the ground plane will appear on the resulting photoprints as displaced images a_1a_1' on photograph 1 and a_2a_2' on photograph 2 (Fig. 3.21).

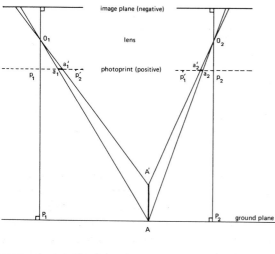

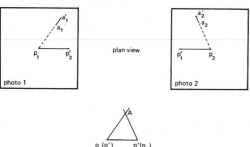

Fig. 3.21   The radial line principle.

By the radial line assumption, the correct plan position, a, on photograph 1 will lie along the radial $p_1a_1a_1'$, and on photograph 2 along the radial $p_2a_2a_2'$. Thus, on each photograph, the direction of the correct plan position, a, of a point from the principal point is known.

The line joining p_1 and p_2' (the image on photograph 1 of the principal point of photograph 2) is the photographic distance equivalent to the air distance moved by the camera between exposures 1 and 2. p_1p_2' is known as the 'photo-base' of photograph 1. On photograph 2, the distance

represents the same air distance as p_1p_2' on photograph 1. Within the limits of slight scale variations from photograph 1 to photograph 2, the photo-bases on the two photographs should be equal in length, since they represent the same line in space.

If a tracing sheet is placed over photograph 1, and the baseline p_1p_2' and also the line from p_1 through a_1a_1' are drawn in, then the correct direction of a from p_1 can be represented graphically. If the tracing sheet is then placed on photograph 2 and the photo-base made to coincide with p_1p_2, then the line from p_2 through a_2a_2' will intersect the existing line on the tracing. The intersection of the radial lines from p_1 and p_2 will then indicate A, the plan position of point a corrected for relief and tilt displacement (Fig. 3.21).

There is a strong similarity between this graphical method of fixing points on aerial photographs and the ground surveying procedure adopted in plane tabling, for fixing the plan position of points by intersection from two ends of a known base line (see Fig. 2.31).

3.6.2.5 THE ARUNDEL METHOD

The procedure described above can be extended from photograph to photograph along a strip, beginning and ending with a photograph in which there are known control points. There should preferably be additional control points in every sixth photograph along the strip. The points which are fixed by intersection, along the edges of a photograph, may then be used as plan control points for the plotting of details later. Extending plan control points in this way is commonly known as the Arundel method (Fig. 3.22).

A long sheet of transparent material, preferably polyester film, is placed over the first photograph, and the photo-base and the rays to the required points along the margins of the photographs are drawn. The transparent material is then registered over the second photograph by making the first photo-base coincide with its homologue on photograph 2. The photo-base common to photographs 2 and 3 is then marked, and the rays from principal point 2 to the required control points on the margins of the photograph are drawn. As the procedure is extended along the strip, each minor control point (mcp) along the margins will be fixed

Fig. 3.22 The Arundel method.

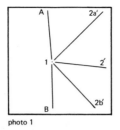

photo 1

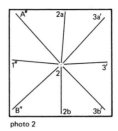

photo 2

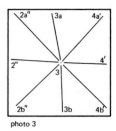

photo 3

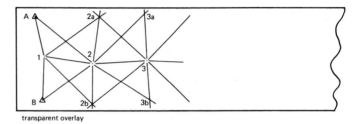

transparent overlay

by a three-ray intersection. Apart from the first and last photographs in the strip, usually nine points per photograph are fixed by this procedure, including the actual principal point and the two transferred principal points on each photograph (Fig. 3.22).

At the end of the strip, the rays drawn to pass through the image of the final control point may not intersect at the known map position of that control point on the transparent sheet. It is then necessary to carry out an adjustment procedure so that the graphical rays through the control points on the strip intersect correctly at the true map positions of these points. The adjustment procedure is described in detail in several other texts but is similar in principle to the method of adjusting a simple traverse in ground surveying (2.2.5.1). The graphical Arundel method is time-consuming, and it can be difficult to carry out the adjustment procedure. The method is partly mechanised and speeded up by the use of 'slotted templates'.

3.6.2.6 SLOTTED TEMPLATE METHOD

The locations of the principal point, the transferred principal points and the minor control points are pricked through each aerial photograph on to a sheet of stable plastic material, preferably a polyester-based or aluminium-based one, and about the same size as the photograph. (A suitable

material is old X-ray film.) By means of a special punch, a circular hole is punched through the principal point on the stable plastic template and, with the help of a special slotted template cutter, the radial rays from the principal point through the other points are replaced by a precisely cut slot, about 5 mm wide and 2 to 6 cm long (Fig. 3.23).

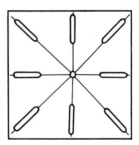

Fig. 3.23 A slotted template.

When a series of these templates has been prepared for a strip of photographs, the templates are laid down, one by one, on a base sheet of stable plastic drawing material that is marked with the known control points, which are fixed by pins to a base board. All slots (rays) passing through the same point are held together by a stud, and, for a correct laydown of the whole assembly, the studs for the control points should be fixed over their respective pins on the base sheet. The slots allow considerable free play in the movement of the whole assembly, so that by expansion or contraction of the template assembly the control point studs can eventually be located over their pins on the base board. In this way the rather tedious graphical adjustment procedure is replaced by a simpler and faster mechanical adjustment.

When the control point pins have had the appropriate studs placed over them, the other studs represent the adjusted positions of the unknown points, and can be pricked through on to the base board.

The slotted template method for extending planimetric control has been widely used in mapping agencies, because of its relative simplicity and the low cost of equipment. Though it is now largely superseded by faster and more precise methods, the slotted template method is still used in countries and in situations where access to sophisticated and expensive equipment is limited.

3.6.2.7 RADIAL LINE PLOTTER

The methods described so far are suitable for correcting relief- and tilt-displacement errors where only a relatively small number of points are concerned. While the graphical method could be applied, point by point, for the correction of the plan position of a boundary or a linear type of feature, the time taken would be unacceptably long for most purposes. The rapid point by point correction for relief and tilt displacements is possible with an instrument know as a radial line plotter, whose design is based on the principle of the Arundel method, which has already been described.

The upper portion of the instrument (Fig. 3.24) consists of a mirror stereoscope which permits stereo-viewing of the aerial photographs which are centred below the large mirrors on two circular metal photo-carriers. When the photographs are set for stereo-viewing, a metal pin is inserted in a transparent perspex cursor, and passes through the principal point of the photograph into a hole in the metal photo-carrier. When centred on the principal point, the perspex cursor is constrained to move radially around that point. The radial index line is usually a fine, coloured thread or an engraved line on the perspex cursor. This cursor is connected to

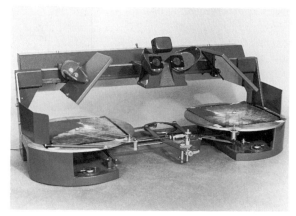

Fig. 3.24 Radial line plotter (Hilger and Watts).

the radial arm, which is located below the circular table.

The radial arm is linked to the plotting bar, which can move in the x and y directions but is constrained by a linkage mechanism to keep parallel with the eye-base. When the plotting bar holding the pencil is moved, the radial index lines rotate in sympathy around the principal point of each photograph. With the continuous rotation of the radial index lines, an almost infinite number of intersection points can be made, but only the actual intersection point need be plotted by the pencil on the plotting bar.

It is necessary to place the control map sheet, or a sheet with four plan control points, on the table below the plotting bar. For most radial line plotters there is a limited range of photograph-to-map scale differences within which the instrument must be used (a common limitation is between 2:3 reduction and 3:2 enlargement). There is a standard setting-up procedure, outlined in the instrument handbook, which ensures that when the radial index lines intersect on the photo-image of a control point, the plotting pencil is over the correct map position of that point.

If no existing map is available to act as a control base, plan control points may be obtained by the radial-line methods of control extension already described (Arundel method or slotted template), or directly by ground surveying methods.

Once the photographs are correctly set up and linked to the control base, the plotting of the required detail features can be carried out, point by point, and at each intersection the relief displacement effect is corrected automatically.

For plotting the detail around the photo-base, where the intersection angle would be very oblique, it is necessary to decentre the perspex cursors and to reposition them at 1 cm above, then below, the principal point. The errors made in this decentring procedure can be shown to be negligible for most radial line plotting work.

3.6.2.8 CORRECTING ERRORS DUE TO TILT DISPLACEMENT

In attempting to correct the planimetric errors in the photo-image caused by tilt displacement it is essential to establish a predictable relation between the ground shape of a feature and its appearance on a tilted photograph. The explanation of this relation lies in the theory of anharmonic ratios (Fig. 3.25).

With reference to Fig. 3.25, if you consider two pencils (a number of coplanar rays with a common

origin) of rays, origins a and A, then the ratio of the intercepts of lines cutting across these pencils of rays is constant, and is the same as the ratio at the intercepts (S_1, S_2, S_3, S_4) of the line at the join of the two pencils of rays.

It can therefore be shown that

$$\frac{S_1 S_3}{S_2 S_3} : \frac{S_1 S_4}{S_2 S_4} = \frac{S_1' S_3'}{S_2' S_3'} : \frac{S_1' S_4'}{S_2' S_4'} = \frac{S_1'' S_3''}{S_2'' S_3''} : \frac{S_1'' S_4''}{S_2'' S_4''}.$$

If the two pencils of rays, origins a and A, are considered to be in the tilted photo-plane and in the ground plane, respectively, then the applicability of this theory to the case of a tilted aerial photograph becomes apparent (Fig. 3.26).

The fact that the anharmonic ratios (cross ratios) in the ground plane are the same as in the plane of the tilted photograph is used in a simple graphical rectification technique to remove the effect of tilt on the plan position of a point on an aerial photograph. Any method designed to correct for plan errors due to tilt is generally called a 'rectification' procedure.

Fig. 3.25 Anharmonic ratios.

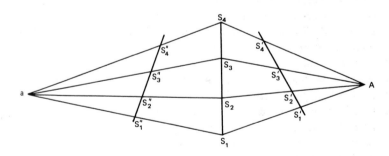

Fig. 3.26 Anharmonic ratios applied to a tilted photograph.

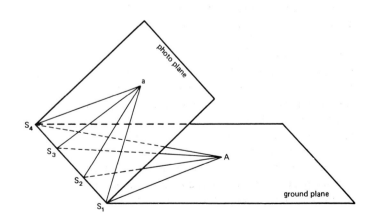

3.6.2.9 PAPER-STRIP METHOD

A practical method embodying the principle of constant anharmonic ratios is the paper-strip method. The correct plan positions of four points must be known (A, B, C, and D on Fig. 3.27), and these points must be identifiable on the photograph (a, b, c and d). The locations of any number of other points may then be transferred from photograph to map, using a paper strip to maintain the same anharmonic ratios on map and photograph.

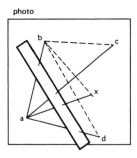

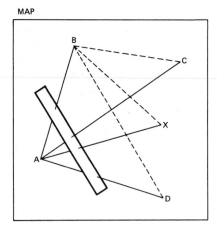

Fig. 3.27 Paper-strip method for transferring point by point.

The procedure is as follows.

1. Using a very fine line, rays are drawn on the photograph and the map from point a(A) to the other three points, and on the photograph to a fourth point x, whose position on the map is required.
2. A paper strip is placed across the pencil of rays from point a on the photograph, as far from it as possible, and is marked where the rays to b, c, d and x cut the edge of the paper strip.

3. The paper strip is then placed over the pencil of rays from A on the map, and the paper strip is moved out and in from A until the marks on the edge of the strip coincide with the rays to B, C and D on the map. The direction of ray AX can then be transferred to the map.
4. Steps 1 to 3 are repeated, using B as the origin of the pencil of rays, and the ray BX can then be drawn in.
5. The intersection of rays AX and BX on the map gives the position of the unknown point X, corrected for the effect of tilt.

By this method, the correct map positions of any number of points may be located from the photographs. The method may be used with near-vertical photographs (tilted up to 6°) or obliques (15°), but ground relief should be less than 5 per cent of flying height. This method should not be used where relief changes rapidly in a small area, but it is satisfactory for rolling terrain. If height differences are considerable, it may be possible to plot the detail in height zones, using the four-point anharmonic method for each of these zones in turn (Fig. 3.28). It would be more correct, however, to use a radial line plotter in such circumstances.

An advantage of this method is that it can be used with almost any type of aerial photography, and knowledge of the focal length or flying height is not required. Moreover, it should be noted that the equipment required is minimal.

3.6.2.10 THE UNION JACK METHOD OF GRAPHICAL RECTIFICATION

The paper-strip method is a time-consuming procedure for the transfer of more than a few

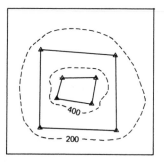

△ plan control point

Fig. 3.28 Plotting in height zones.

points from photograph to map, so that for the transfer of a greater amount of plan detail a method such as the Union Jack or Flag method is preferable, since only two points need to be fixed by the paper strip. The procedure is as follows:

1. The basic requirement is that the positions of four points should be known, on photograph and map. These points should be marked and the quadrilateral they form constructed. The diagonals of the quadrilateral are then drawn in (Fig. 3.29).

photo

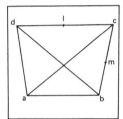

MAP

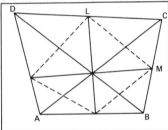

final subdivision of grid

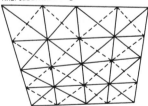

Fig. 3.29 Union Jack method.

2. By means of the paper-strip method, with point a as origin, points l and m are transferred from photograph to map. Points l and m are chosen to be the approximate mid-points of sides cd and bc on the photograph (they may or may not coincide with points of natural detail). The accuracy of the final map depends largely on the

correct transference of l and m from photograph to map.

3. The lines from L and M, through the intersection of the diagonals, to AB and AD are drawn in.

4. The figure then consists of four small quadrilaterals with one diagonal in each. The second diagonal is then added to each of these four small quadrilaterals.

5. The lines joining the intersections of the diagonals of the four small quadrilaterals can be drawn, and the subdivision of the figure, on photograph and map, can proceed, retaining the Union Jack figure at each stage of the subdivision (Fig. 3.29).

6. The extent of the subdivision of the figure depends on the amount of detail to be transferred. The detail is transferred by visual comparison, one triangle at a time, keeping the same proportions within corresponding triangles on map and photograph. The sort of details that might be transferred are the drainage system, vegetation and soil boundaries, and other specialist information, for these may be visible on the photograph but are generally absent from published topographic maps. The net result of this procedure is to transfer detail from the photograph to the map base, correcting for the tilt displacements and for the scale difference between photograph and map.

Several optical instruments are available which are designed to fulfil the task of graphical rectification already described.

3.6.2.11 THE SKETCHMASTER

The graphical rectification procedure is simplified by an optical instrument designed to speed up the mapping process. This instrument, the Sketchmaster, uses single aerial photographs (Fig. 3.30).

The photograph is fixed to the photo-carrier and the Sketchmaster is placed over a map of the area of interest. (In the absence of an existing map, a framework of four control points must be provided.) By adjusting the ratio of the distances, eye-to-photograph and eye-to-map, it is possible to compensate for the scale difference between one and the other. With most instruments a range from

Fig. 3.30 Zeiss Aero-Sketchmaster.

0.5× reduction to 2.5× enlargement is possible between photograph and map.

The photo-carrier can be tilted to correct for the tilt errors present in the photograph. With the Vertical Sketchmaster (Fig. 3.31), adjustment of the relative settings of the footscrews may be sufficient for small tilts. This adjustment may be checked by making photographic and map detail of the same features coincide before proceeding to add the required new information to the map base. The coincidence of map and photographic detail is observed by means of a double-prism device or a

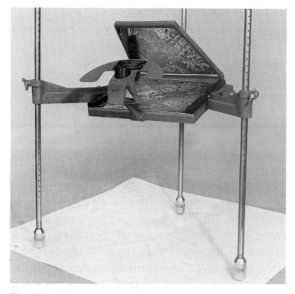

Fig. 3.31 Nashton Vertical Sketchmaster (Type S-14).

semi-reflecting mirror in the Vertical Sketchmaster, which makes it possible for the observer to view the map and photograph simultaneously (Fig. 3.30). It is usually necessary to try various lighting arrangements for map and photograph in order to achieve a proper balance of illumination between the white of the map sheet and the grey tones of the photograph. When the images of map and photographic detail are in coincidence, fresh detail can be added to the map base by following the outline of the required features with a pencil.

The setting-up procedure may be simplified if similar grids are drawn on map and photograph, as in the Union Jack method. The grid lines may then be made to coincide for a correct setting of the instrument. Experienced users, however, seldom find it necessary to adopt this procedure.

3.6.2.12 MAPPING FROM LOW OBLIQUE PHOTOGRAPHY

In addition to correcting for the tilt errors present in near-vertical photographs (up to about 6° from the vertical), some Sketchmasters are designed to cope with the plotting of plan detail from low oblique photographs, where the camera axis has been intentionally set at 15° or 20° from the vertical. An example of such an instrument is the Nashton Vertical Sketchmaster Type S-14 (Fig. 3.31), which may be used for planimetric mapping from 'fan' photography.

'Fan' photography results from a multi-camera set-up in the aircraft, where the optical axes of the cameras are tilted to permit photographic coverage of as much of the ground area as possible (Fig. 3.32). This was quite a common procedure for much of the aerial photography of Great Britain done by the Royal Air Force. A great deal of this work is in circulation and is used by many field scientists. For many parts of Great Britain, in fact, RAF 'fan' photography may be the only available aerial source for the 1940s and 1950s. For this reason it is a valuable period record of the landscape, especially for field scientists who may be interested in recording changes over a time span of several decades.

The 'fan' photographs are normally identified by the letter F, followed by a two-digit number in the title strip of the photograph: the first digit indicates

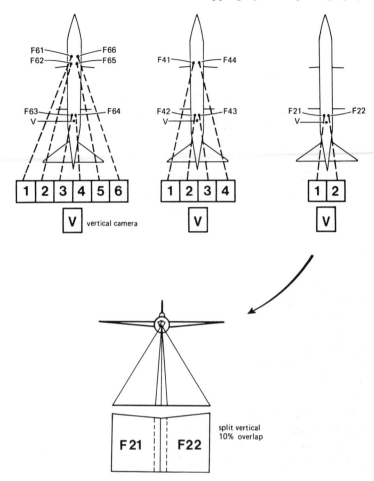

Fig. 3.32 Various camera configurations giving 'fan' photography.

the number of cameras used in the fan, and the second the camera position, numbered from port to starboard. The outer photographs of wide fans are very oblique, whereas the innermost pair, sometimes known as 'split verticals' or 'twin-obliques', may be nearly vertical.

The normal tilt of 'fan' photography can be deduced from the reference numbers. For example, with a 20 inch (51 cm) focal length lens, F41, F42, F43 and F44 (inner photographs) have a nominal tilt of 10° (see Fig. 3.32). When the photograph is placed on the photo-carrier of the Sketchmaster, the carrier should be set, on the angle scale, to a tilt of 10°, before any attempt is made to adjust the legs or to match photographic and map detail.

At low angles of tilt (up to 10°) the initial height settings can be the same as for vertical photography, with roughly the same scale factor, but more adjustment may subsequently be required. If a

general fit cannot be obtained, the only way to proceed is to make the detail on the photograph coincide with the map, one part at a time, resetting the instrument on moving to the next area.

The aim of the setting-up procedure is to achieve a 'best-fit' of photographic detail to corresponding map detail before transferring, by eye, the additional information required on the map base. The smaller the tilts and the flatter the ground area, the easier it will be to effect coincidence of detail on photograph and map.

With more rugged ground, the relief component of image displacement becomes more appreciable; the Sketchmaster then becomes less suitable for mapping, since it employs single photographs, and cannot correct the displacements arising from variations in ground height. If a Sketchmaster is the only instrument available, however, and provided that the ground relief is not excessive (less than 5 per cent of flying height), it is possible to minimise the effect of relief displacements by using the central portion of every photograph in a strip of photographs. This will, of course, require a great deal of time to perform and it may lead the user to seek a more efficient procedure.

3.6.2.13 THE STEREOSKETCH

A more sophisticated and expensive variant of the Sketchmaster is the Stereosketch, originally manufactured by Hilger and Watts Ltd. The major advantage of this instrument is that it permits stereo-viewing of the photographs, which is more comfortable for the user and also improves the interpretability (Fig. 3.33).

The drawing table (base map) can be moved vertically towards and away from the viewing eyepieces, thus altering the magnification ratio between photograph and map (in the range 0.7× to 2.35×).

Instead of tilting the photo-carrier (as in the Sketchmaster) the tilt is applied to the base map, (± 5°) in the direction of flight and at right angles to that direction. The instrument is therefore only suitable for plotting plan detail from tilted near-vertical photographs, and could not cope with 'fan' photographs with angles of tilt of 10° or 20°.

Although the upper portion of the Stereosketch superficially resembles a radial line plotter, the two instruments should not be confused, since the Stereosketch cannot remove the effects of displacements due to ground relief.

3.6.2.14 THE ZOOM TRANSFER SCOPE

Another instrument designed to speed up the transfer of plan details from aerial photograph (or satellite image) to map base is the Zoom Transfer scope (Fig. 3.34). The photograph is placed on the horizontal glass stage and the map on the table below the map lens. A zoom magnification control knob permits the photo-image and the map to be viewed simultaneously at the same scale. Image distortions due to camera tilt can be removed, or at least minimised, by the use of the 'image stretch' lever which permits differential stretching of the image in two directions at right angles. Once details common to photograph and map have been registered, the operator can begin the transfer of other detail from photograph to map base. With appropriate map lens selection a scale enlargement of up to 14 times can be achieved between photograph and map base, so that in addition to its traditional use in map revision from aerial photography this

Fig. 3.33 The Stereosketch (Hilger and Watts).

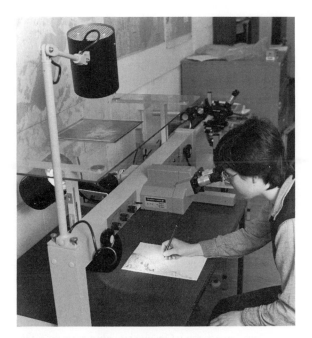

Fig. 3.34   Zoom Transfer Scope (Bausch and Lomb).

instrument is also suitable for 1:100 000 scale mapping of plan detail from 1:1 000 000 scale LANDSAT images (see sect. 4.6.3.1)

3.7 HOW TO OBTAIN HEIGHT VALUES FROM AERIAL PHOTOGRAPHY

3.7.1 METHODS EMPLOYING SINGLE PHOTOGRAPHS

On near-vertical aerial photography the height of an object may be obtained if the vertical dimension in the photo-image or the shadow of the object is clearly visible on the photograph. The methods of obtaining height values from oblique photography are more complex. A detailed treatment may be found in the *Manual of Photogrammetry* (Slama, ed. 1980).

3.7.1.1 HEIGHT FROM RELIEF DISPLACEMENT
(Figs. 3.18 and 3.19)

If the top and base of a vertical object are clearly visible on the aerial photograph, the height of the object causing the relief displacement of the image can be deduced from the formula

$$\Delta h = (H - h)\ \frac{de}{r}$$

where: $(H - h)$ = the ground clearance of the aircraft,

de = measured radial displacement of the image of the top from the image of the base of the feature.

r = measured radial distance from the principal point to the image of the top of the feature.

3.7.1.2 HEIGHT FROM SHADOW LENGTH

The top and base of a vertical object may not always be visible on an aerial photograph, but the shadow cast by the object can often be seen clearly. The shadow lengths on the photograph are in proportion to the height of the object casting the shadow, although correction may be required for ground slope.

If the shadow of an object of known height is measured, heights of objects with any other measured shadow lengths may be deduced. If there is no object of known height in the photograph, it is still possible to deduce heights from shadow lengths, provided that the time and date of flight and the latitude of the place are known. This is possible because there is a relation between height of object, length of shadow and altitude of the sun at the time of photography (Fig. 3.35). The sun's altitude for a place at a particular time is obtained from astronomical tables, but the method is often laborious. A full treatment of the method is given in Howard (1970).

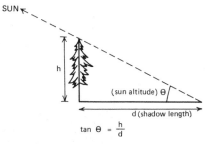

Fig. 3.35   Relation between object height, shadow length and sun altitude.

3.7.2 METHODS EMPLOYING STEREOSCOPIC PHOTOGRAPHS

If aerial photographs are taken in an overlapping sequence (Fig. 3.15), the simultaneous viewing of a pair of successive photographs will provide a three-dimensional image of the ground in the area of the overlap. The possibility of obtaining a measure of the height dimension perceived in the stereoscopic model is of fundamental importance in topographic mapping from aerial photographs.

Before heights can be derived from stereophotography, it is necessary to establish a mathematical relation between some measurable parameter on the photograph and the true ground height of the object. This relation is most clearly appreciated by considering the stereoscopic appearance of a simple feature, such as a dot, rather than of the complex images on a photograph (Fig. 3.36).

In plan view, the dots 1′ and 1″ are closer together than are 2′ and 2″, which, in turn, are closer together than are 3′ and 3″. When the pairs of dots are examined with a simple lens stereoscope, the dot formed by the stereoscopic fusion of dots 1′ and 1″ appears to be higher above the page than the dot formed by fusing 3′, and 3″, and the fused dot 2′, 2″ appears to be floating at a level between the other two. It would appear, therefore, that the apparent height of the fused or 'floating' dot is related to the horizontal separation of the dot images.

In cross-section (Fig. 3.36) the pairs of dots are shown as if recorded on aerial photographs, and it can be seen that the height differences of the objects is related to the differences in spacing of the images. This observation is true only if the images are observed as if on a line parallel to the observer's eye-base. On aerial photographs, this means that the height differences are related to image separations measured along a line parallel to the photo-base (the line joining the two principal points in the overlap area). When two aerial photographs are set up for correct stereo-viewing the photo-base will be parallel to the eye-base of the observer. The photo-base is a fundamental line in stereo-photography, and it is used as the x-axis of the aerial photographic coordinate system, with origin at the principal point of each photograph (Fig. 3.37).

The x-axis corresponds to the direction of flight, and the y-axis is along the direction of the wingspan of the aircraft. With a photo-coordinate system it is possible to give a unique numerical reference for the location of any point on a photograph.

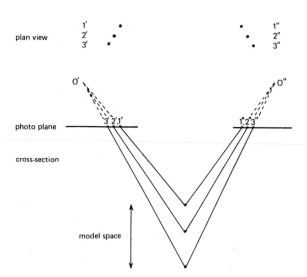

Fig. 3.36 Stereoscopic perception of a dot.

Fig. 3.37 Photo-coordinate system (ideal).

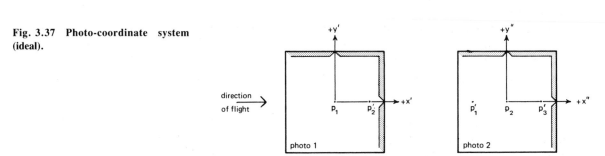

Note that $P_1 P_2'$ may not be parallel to the edge of the photograph if the aircraft has deviated from the projected straight course, due to drift or crab.

Fig. 3.38 Parallax and apparent image shift.

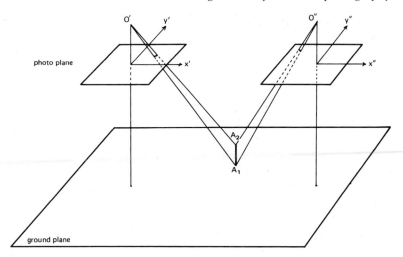

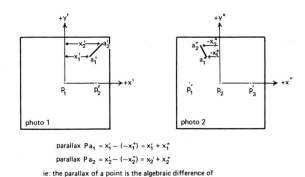

parallax of a_2 is the apparent x-shift from photo 1 to photo 2

3.7.2.1 PARALLAX

In recording the image of a ground object in the overlap portion of two successive aerial photographs, the location of the image appears to change from photograph 1 to photograph 2. This change in location may be appreciated by considering the images with reference to the photo-coordinate axes (Fig. 3.38). In the first photograph, the image may be recorded in the top right quadrant of the photograph, but in photograph 2 the image is in the top left quadrant. This image shift is due to an effect known as parallax, which is defined as 'the apparent shift in the position of an image due to a change in the viewing position'. In aerial photography the 'change in the viewing position' is provided by taking the photographs from two different camera positions. When a pair of photographs are correctly positioned for stereo-viewing, the differences in the image locations of a point are confined to the x-direction. (If corresponding images differ in the y-direction, stereo-fusion of the images may be achieved, but with some eyestrain, or may not be possible at all.)

If the x-coordinates of an image are measured on photograph 1 and photograph 2, the algebraic difference of the x-coordinates (obtained by subtraction, taking note of the sign) will give a measure of the apparent image shift of a point, i.e. the parallax of the point (Fig.3.39). If precise

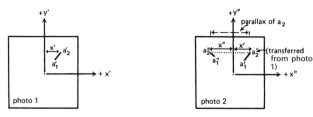

parallax $Pa_1 = x'_1 - (-x''_1) = x'_1 + x''_1$

parallax $Pa_2 = x'_2 - (-x''_2) = x'_2 + x''_2$

ie: the parallax of a point is the algebraic difference of the x-coordinates on photo 1 and photo 2

Fig. 3.39 Parallax and photo-coordinates.

coordinate measurements were made for several points in the overlap area, it would become apparent that the parallax values are related to the heights of the points, and since the y-coordinate differences are zero for correct stereo-viewing, only x-parallaxes need be considered.

The relation, noted earlier, between the heights of objects and the linear separation of the images in stereo-photography can, therefore, be expressed more specifically as being due to differences in the x-parallaxes of the objects, and is given numerically by the algebraic difference in the x-coordinates of the images on successive photographs. For vertical aerial photographs taken from the same flying height, points at the same height in the terrain will have the same parallax values.

3.7.2.2 PARALLAX FORMULA

The relation between height differences and parallax differences is normally given by the following mathematical expression (see Fig. 3.40).

$$\Delta h = Z_1 \frac{\Delta p}{b + \Delta p}$$

where: Δh = height difference between principal point 1 and any other point n;

Z_1 = flying height over principal point 1;

b = photo-base (the parallax of principal point 1); and

Δp = parallax difference between principal point 1 and point n.

The general formula gives the height difference between any two points, where one of the points is the reference point. There are difficulties in the direct measurement of the parallax of the reference point, since it involves setting out the coordinate axes on the two photographs, measuring the x-coordinates of the point and then taking the algebraic difference of the measured values. If the

Fig. 3.40 Relation between parallax differences Δp and height differences Δh.

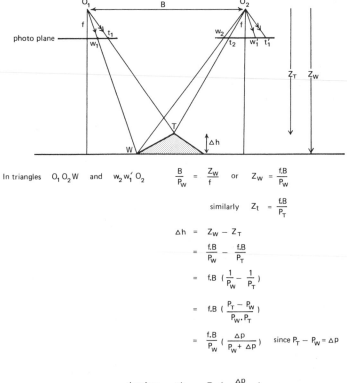

In triangles $O_1 O_2 W$ and $w_2 w_1' O_2$ $\dfrac{B}{P_W} = \dfrac{Z_W}{f}$ or $Z_W = \dfrac{f.B}{P_W}$

similarly $Z_t = \dfrac{f.B}{P_T}$

$$\Delta h = Z_W - Z_T$$

$$= \frac{f.B}{P_W} - \frac{f.B}{P_T}$$

$$= f.B \left(\frac{1}{P_W} - \frac{1}{P_T} \right)$$

$$= f.B \left(\frac{P_T - P_W}{P_W . P_T} \right)$$

$$= \frac{f.B}{P_W} \left(\frac{\Delta p}{P_W + \Delta p} \right) \quad \text{since } P_T - P_W = \Delta p$$

therefore $\Delta h = Z_W \left(\dfrac{\Delta p}{P_W + \Delta p} \right)$

This gives the height difference between W and T with W as a reference level left-hand principal point is taken as reference, then the x-coordinates of that point are O, on photograph 1, and −b (the photo-base) on photograph 2. The algebraic difference gives the right-hand photo-base b as the parallax of the left-hand principal point. This is a single measurement, which can be made without the need to construct coordinate axes on the photographs.

3.7.2.3 MEASUREMENT OF PARALLAX DIFFERENCE Δp

The parallax difference Δp is proportional to the algebraic difference of the x-coordinates of a point (a linear measurement), but the parallax differences are usually so small (a few millimetres or tenths of a millimetre) that direct linear measurement on the photograph is not sufficiently precise. If, however, the photographs are viewed stereoscopically, the human eyes can detect depth changes within the stereo-model equivalent to linear x-changes in the photograph of a few hundredths of a millimetre. For this reason the parallax differences Δp are determined by a method employing stereoscopic viewing.

When a mirror stereoscope is used for stereo-viewing, the instrument used for measuring parallax differences is a parallax bar, or stereometer (Figs. 3.41 and 42). The parallax bar consists of two thin glass plates, each fixed to a metal tube; one tube slides smoothly into the other, thus allowing the separation of the glass plates to be altered. A small mark, such as a dot, is etched on each glass plate, and the precise separation of the dots is determined by a micrometer screw, which controls the movement of one tube inside the other. The scale on the micrometer screw reads directly to 0.05 mm and to 0.01 mm by interpolation.

When an overlapping pair of aerial photographs is set up for stereo-viewing, the left-hand glass plate is positioned so that the dot covers the image of the reference point on the left-hand photograph. The micrometer screw is then turned until the dot on the right-hand plate moves to cover the image of the same point on the right-hand photograph. When viewed stereoscopically, the mark formed by the fusion of the two dots appears to rest on the surface of the stereo-model at the reference point. Moving the dots slightly closer together, by means

Fig. 3.41   Mirror stereoscope and parallax bar.

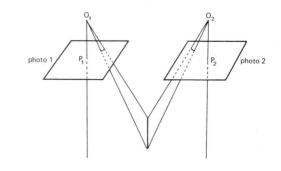

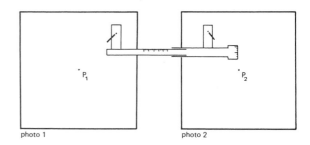

Fig. 3.42   Measurement of Δp with a parallax bar.

of the micrometer screw, makes the fused mark appear to rise within the model and to form a 'floating mark' in the model space; moving the dots apart again causes the floating mark to appear to drop on to the reference point. The human eyes are very sensitive to the positioning of the floating mark within the model, and it is possible to detect changes in the height of the floating mark equivalent to linear x-differences of a few hundredths of a millimetre. The parallax bar reading for the reference point may then be recorded. This reading is of no particular significance in itself – it is not the value of the parallax of the point.

The separation of the dots is then altered until the 'floating mark' appears to rest on the surface of the other point. In Fig. 3.42, for simplicity, the measurements are made at the base and at the top of a chimney. The base of the chimney is about the same elevation as principal point 1. The difference between parallax bar readings 1 and 2 is the value Δp in the parallax formula.

The method described is satisfactory for the determination of height differences where the two points concerned are close together in the photograph, e.g. the top and base of a chimney, a tower or a tree. If the method is applied for the determination of a large number of spot heights, with a view to preparing a contour map, the heights obtained from the formula will contain large errors, since the parallax bar measurements are made with the photographs placed flat on the table. No allowance is made for the fact that the stereo-model has been formed from tilted photographs, so that the measurements are actually being made on a deformed model surface. If the reference point and other point n are close together on the photograph, the deformations will have little effect, but if points are widely separated on the photograph, the full effects of the deformations will be present in the computed heights. If a large number of spot heights are required for contouring, the computed values of Δh provide only 'crude heights' that must be corrected for the model deformations (Fig. 3.43). Several procedures have been developed for the correction of crude heights to true heights, by means of correction formulae based on the height errors at a number of points of known height. (The method devised by Thompson is described in detail in the *Photogrammetric Record*, 1954.) While the equipment used in this heighting procedure is fairly simple (mirror stereoscope and parallax bar), the measurements and mathematical treatment to correct the results can be a laborious exercise. The effort of obtaining a large number of spot heights for contouring a stereo-model can be greatly reduced if the user has access to an electronic computer, which allows the rapid solution of the mathematical equations. The use of parallax bar and mirror stereoscope with an electronic computer to solve the correction equations, is described by Methley (1970).

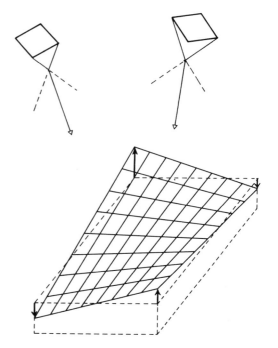

Fig. 3.43 Model deformations that influence height determination.

The best heighting results achieved in Methley's experiment were spot heights to an accuracy of 0.025 per cent of flying height. This compares favourably with the heighting accuracies of photogrammetric plotting instruments (the range is 0.01 to 0.05 per cent of flying height), and at only a fraction of the cost. The best results, however, are only achieved with reliable parallax bar readings, and this is possible only after the observer has had considerable practice.

Many people use aerial photographs without having access to expensive plotting instruments to carry out the mapping. With an understanding of the theoretical background of model deformations and correction formulae, and with a minimum of equipment, it is possible to obtain spot heights to an accuracy approaching that of many stereo-plotting instruments.

A detailed treatment of the theory underlying the construction of contour maps from spot heights obtained by parallax bar measurements can be found in most textbooks on photogrammetry.

3.7.2.4 LENS STEREOSCOPE AND PARALLAX WEDGE

The standard parallax bar cannot be used with a

Fig. 3.44 Zeiss TM pocket measurement stereoscope.

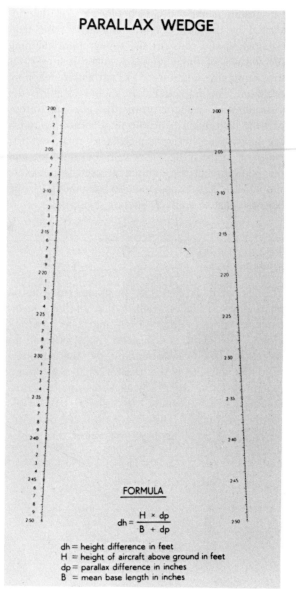

PARALLAX WEDGE

FORMULA

$$dh = \frac{H \times dp}{B + dp}$$

dh = height difference in feet
H = height of aircraft above ground in feet
dp = parallax difference in inches
B = mean base length in inches

Fig. 3.45 The parallax wedge.

pocket or lens stereoscope. One type of lens stereoscope has a small stereometer attachment, similar to a parallax bar, incorporated in a base frame (Fig. 3.44). A more common aid for parallax heighting by lens stereoscope is the parallax wedge, which is a simple analogue of the parallax bar. Instead of varying the dot separation continuously with a micrometer screw, however, the dot separations are set out at predetermined intervals. The wedge is a rectangular piece of stable transparent material marked with two diverging lines of black dots. The separation of corresponding pairs of dots varies from about 50 mm to 65 mm, in steps of 0.05 mm. The wedge is placed on top of the photographs when viewing in stereo, with one line of dots on the left photograph and the other line on the right photograph. When corresponding pairs of dots are fused stereoscopically, the observer should see an inclined line of floating dots in the model space. The observer's task is to judge which dot in the floating line is making contact with the ground surface in the model at the point of interest. Points that are high in the model space will require a pair of dots near the converging end of the wedge (Fig. 3.45).

3.8 MAPPING HEIGHT AND PLAN IN ONE OPERATION

Where the highest accuracy is required in preparing a map from stereoscopic aerial photographs, it is necessary to use a technique that corrects simultaneously for the relief- and tilt-displacement errors present on near-vertical aerial photographs. The equipment used for this purpose is a stereo-plotting instrument, which can produce a direct plot of plan detail and contours in a single setting.

Stereo-plotting instruments are expensive, and are thus not so readily to hand as other equipment discussed in this book; for example, a stereo-plotter for medium-scale topographic mapping may cost about twenty times as much as a theodolite suitable for ground surveys by tacheometry. As demonstrated by the flow diagram in Fig. 1.2, the field scientist should consider the topographic mapping possibilities of this equipment, since it may offer the optimum solution to a particular mapping problem. The potential user should establish the location and availability of the nearest stereo-plotting instrument, since he may be able to enlist the help of a photogrammetrist to carry out the mapping. If the project is of sufficient duration and the photogrammetric equipment used is relatively simple, it might be possible and reasonable for the field scientist to learn to operate it himself.

3.8.1.1 WHAT IS A STEREO-PLOTTING INSTRUMENT?

The essential problem in mapping from aerial photographs is the transformation of the coordinates of photographic prints into their correct ground (or map) coordinates. It is possible to represent this transformation by mathematical formulae, and this is the basis of 'analytical photogrammetry', but in practical terms the transformation is usually achieved by using a stereoscopic plotting instrument. The stereoplotter can therefore be regarded as an analogue computer that solves

continuously the mathematical equations linking the photo-coordinates and the ground-coordinates of points in the stereo-model.

The stereo-plotters that most closely satisfy the mathematical relations between photo-coordinates and ground-coordinates are called 'exact-solution' instruments. Those that only partly fulfil the mathematical relations are 'approximate-solution' instruments.

3.8.1.2 PORRO–KOPPE PRINCIPLE

In the early days of photogrammetric mapping the image displacements produced by the lens distortion of the aerial camera were an additional error source that had to be corrected. The portion of the stereo-plotter that simulates the aerial camera is usually termed the projector, and one way of removing lens distortion effects is to re-project the photographs of a stereo-pair through two projectors with lens characteristics identical to those of the aerial camera; this method of re-projection, which is known as the Porro–Koppe principle, was fundamental to the design of many of the early stereo-plotting instruments.

One of the earliest and simplest designs of plotting instrument based on the optical projection

Fig. 3.46 Multiplex plotting machine.

principle was the Multiplex, in which the geometry of the projector is a reduced-scale version of the camera used to take the photographs. The projectors are therefore small enough to permit several to be positioned on one projector bar, each one simulating a different camera position in the progress of the aircraft along the flight track (Fig. 3.46). Although less accurate than the plotting instruments which use full-size photographs, the Multiplex was, in the 1930s and 1940s, an extremely popular instrument, with many hundreds of them being used in the basic topographic mapping of the USA at that period. One of the major disadvantages is that work must take place standing in a dark-room with the operator viewing through blue and red spectacles the pair of projected images forming a stereoscopic image. Operator discomfort was a major factor in the demise of the Multiplex. In terms of availability and cost, however, it offers considerable possibilities for use by non-specialist operators.

Most modern aerial survey cameras have lenses with negligible distortion characteristics (less than 10 μ), so that one of the original attractions of the Porro–Koppe design is no longer so important. This has led to the development of instruments which are based on the mechanical projection principle, whereby the light rays connecting an image point to a model point are simulated by a pair of mechanical space rods (Fig. 3.47). With this type of instrument the analogy between the aerial camera configuration and the plotting instrument is not so readily apparent as with the optical projection type of instrument.

3.8.2 FORMATION OF THE STEREO-MODEL

The best way to appreciate what is happening in a stereo-plotter is to consider the way in which the three-dimensional shape of the ground is recorded in the overlap area of a pair of aerial photographs (Fig. 3.48).

3.8.2.1 INNER ORIENTATION

In order that the bundle of rays emanating from each projector of the plotting instrument may be congruent with the bundle of rays entering the camera lens at exposure, certain geometric characteristics of the aerial camera must be duplicated in the plotting instrument. These camera characteristics are known as the 'inner orientation'. The focal length of the camera lens and the location of the principal point are two fundamental elements in the formation of the bundle of rays. In the plot-

Fig. 3.47 Mechanical projection plotter.

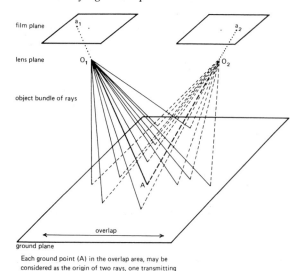

film plane

lens plane

object bundle of rays

overlap

ground plane

Each ground point (A) in the overlap area, may be considered as the origin of two rays, one transmitting an image to photo 1 (a_1) and the other producing an image on photo 2 (a_2)

The almost infinite number of rays from the ground, passing through the lens centre, is known as a 'bundle of rays'.

Fig. 3.48   Recording the stereo-model area.

ting instrument the photographs are correctly centred, in order to locate the principal point, and the principal distance scale is set to the value of the calibrated focal length of the camera taking the photographs (Fig. 3.49).

3.8.2.2 OUTER ORIENTATION

The location in space of the camera with respect to a coordinate system based on a ground reference (e.g. horizontal and vertical axes) forms part of the 'outer orientation', which may be divided into relative orientation and absolute orientation.

Relative orientation

When a pair of overlapping photographs is inserted into a plotting instrument of the optical-projector type (Fig. 3.50), the angular positions of the photographs must be adjusted until the re-projected pairs of rays, from left and right photographs, all intersect at the surface of the optical model of the terrain. This procedure is called 'setting the relative orientation' of the two photographs.

Absolute orientation

When the two projectors of the plotting instrument

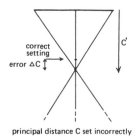

correct setting

error ΔC

C'

principal distance C set incorrectly

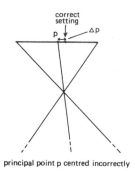

correct setting

p Δp

principal point p centred incorrectly

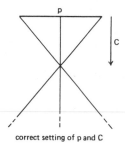

p

C

correct setting of p and C

Fig. 3.49   Inner orientation.

have been set in their correct relative angular positions, an optical model of the terrain may be viewed. Before the model can be measured, it is necessary to establish its size (scaling) and its orientation in space with respect to the height datum (levelling). It is not possible to proceed with mapping unless the correct ground (or map) coordinates, including heights of a number of points, are known. Photogrammetric mapping still requires a link with the ground, through an existing map or by ground surveying for control points.

Scaling

This entails checking a measured distance in the optical model (location of two points) against the known true locations of the points, either on an existing map of suitable scale, or on a gridded sheet on which points determined by ground surveying

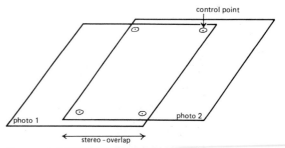

If the above points are sharp detail features, they may also serve as plan control. It may be necessary, however, to select additional plan control points near to the indicated height control points.

Fig. 3.51 Preferred locations of height control points in a stereo-model.

floating dot while viewing the stereo-model. The levelling procedure is complete when the floating dot, placed on each of the four control points in turn, produces a reading of the true height of each point on the height scale of the stereo-plotter. If the height scale is then set at a particular value, say 10 m, and the operator moves the floating mark around the model, keeping it in contact with the terrain surface, the plotting pencil will trace the 10m contour line on the map sheet.

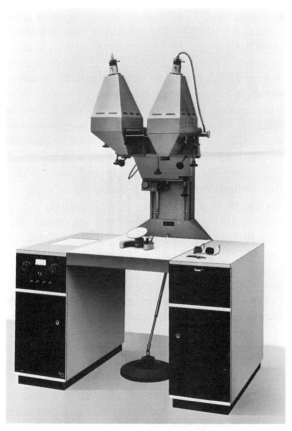

Fig. 3.50 Zeiss DP-1 stereo plotting instrument.

have been plotted. The minimum number of points is two, as far apart as possible, but in practice three or more points are preferred in order to provide a check.

Levelling

After scaling, the model may be viewed at a known scale, but the accurate plotting of detail and the determination of heights cannot proceed until the model has been correctly orientated with respect to the height datum (in Great Britain this is mean sea level at Newlyn, known as the Ordnance Datum or OD). This procedure is known as levelling the model, and depends on prior knowledge of the true heights (referred to OD) of a number of points, preferably a point in each corner of the stereo-model (Fig. 3.51). Every stereo-plotter has a viewing system whose purpose is to allow the observer to see the stereo-model and the measuring mark (floating dot) that is used to follow the detail in the model. It is possible to raise and lower the

3.8.3 FIELD IDENTIFICATION

The operator, having established the correct scale and correctly levelled the stereo-model, is able to begin plotting the required detail on to the map base or the control sheet. Except where very simple features are to be mapped, it is preferable for the operator to establish what is to go on the map by a separate interpretation phase. The value of such a phase is recognised by several national topographic mapping agencies, and a detailed consideration of the benefits of a separate topographic interpretation phase has been given by Tait (1970).

The field scientist will often be recording information of a highly specialised nature on his maps, and the necessity for a separate interpretation phase, by the specialist himself, is usually essential, since the operator of the plotting instrument is unlikely to have any knowledge of the specialised distributions. The interpretation phase is most often carried out by marking the distributions on transparent overlays (placed over the photographs during the interpretation). These overlays are then registered on the photographs in the plotting instrument, and the operator can then easily follow the lines of the specialist's interpretation.

3.8.3.1 INTERPRETATION BEFORE PLOTTING

The user of aerial photography is seldom likely to carry out complete mapping of the required detail without visiting the area. For topographic mapping, visits to the field area are for two main purposes.

1. To add, by ground surveying methods, required details that are not visible on the aerial photograph (e.g. detail hidden by trees or buildings or in dark shadow).
2. To identify features that are not clear from the photo-image alone.

The reliability of the field scientist's interpretation phase can be greatly improved by checking the photographic detail and tones against the situation on the ground in selected areas. This procedure is of assistance in preparing a key for the interpretation. If a field check is carried out after the interpretation phase, it is also possible to check the accuracy of the laboratory interpretation.

3.8.4 WHEN IS PHOTOGRAMMETRIC MAPPING APPROPRIATE?

While, in theory, the photogrammetric and the ground surveying methods of mapping may be considered as alternatives, in practice one method is sometimes preferable to the other. Photogrammetric mapping should be the preferred solution in the following situations.

Where the survey area is very extensive
Extensive generally means from a square km upwards, depending on the detail required. The time-saving advantage of aerial survey becomes more apparent with increasing area.

Where access to the survey area is difficult
This could be either because of remoteness (e.g. if it lies hundreds of kms from home base) or

because of the difficult physical nature of the terrain for ground surveying (e.g. on inter-tidal mudflats).

Where a high density of detail, especially linear detail, is required from the survey
This applies, in particular, to the demarcation by lines of the boundaries of plant communities, for vegetation mapping; to establish these boundaries by ground surveying methods is very time-consuming, even for a relatively small area. Where the communities are clearly identified by tone and texture on a photograph, it is a relatively simple matter, once the photographs are set up for plotting, to follow the outlines of the plant communities on the photograph.

The photogrammetric method has the great advantage that it permits continuous plotting of linear elements, whereas ground surveying techniques can only fix a line on a point-by-point basis.

Where repetitive aerial photographic coverage is of value
The taking of several sets of photographs of the area at different times permits the mapping of changes. Where significant changes in morphology or distributions have occurred over a long time-span, it may be possible to recover and map the situation of 20 years ago by using photographs from archives.

3.8.5 TERRESTRIAL PHOTOGRAMMETRY

Terrestrial photogrammetry is a blanket term which is often used to describe the use of photogrammetric techniques in a situation where the camera is connected to the ground rather than suspended in free space on an airborne platform. The camera is most commonly fixed to a tripod, and the optical axis is in a horizontal plane. If the area of interest lies in a vertical plane (e.g. the face of a building or a vertical cliff), it may be possible to map the details of the object with reference to a vertical plane. This is simply the aerial photographic situation rotated through 90°. If the variation in depth in the object is small compared with the distance from camera to object, it may be possible to map the details by means of a conventional plotting instrument designed for mapping from near-vertical aerial photographs.

If the depth range in the object is more than about 20 per cent of the camera-to-object distance, special plotting equipment suitable for terrestrial photographs must be used. This equipment must always be used for plotting with reference to a horizontal plane when the photographs have been taken with the camera axis horizontal.

It is also possible in terrestrial photogrammetry to use the camera in the vertical mode, where the object of interest is small and lies mainly in a horizontal location. The plotting from such photography is then analogous to the case of near-vertical aerial photography. This method has been used in the mapping of micro-landforms on the earth's surface, e.g. to measure the changes in the form of small gullies after a period of heavy rain (Lo and Wong, 1973).

Terrestrial photogrammetry is preferable to ground surveying methods when it is important not to disturb the object being measured. Instances include mapping the form of sand dunes, or other landforms with relatively unstable slopes or delicate features, where to place a theodolite or a surveyor's staff on the feature would disturb the object being mapped.

Terrestrial photogrammetry is not a direct alternative to ground surveying, but it offers a solution in many instances to mapping problems where ground surveying methods are inappropriate. Recording the images of the scene on photographs also means that the process of measurement can be postponed, since the scene can be recovered from the photographs at a later date.

FURTHER READING

Colwell, R. N. (ed.), 1960, *Manual of Photographic Interpretation*. American Society of Photogrammetry, Falls Church, Virginia, USA.

Dickinson, G. C., 1979, *Maps and Air Photographs* (2nd edn). Edward Arnold.

Kilford, W. K., 1979, *Elementary Air Survey* (4th rev. edn). Pitman.

Slama, C. C. (editor-in-chief), 1980, *Manual of Photogrammetry* (4th rev. edn). American Society of Photogrammetry, Falls Church, Virginia, USA.

Chapter 4 REMOTE SENSING TECHNIQUES

4.1 INTRODUCTION

'Remote Sensing' is a term first coined in the 1960s to describe any method whereby data or information relating to an object are obtained by some sensing device held at a distance from the object. Therefore, in the context of imaging the earth's surface, aerial photography is a form of remote sensing. Data relating to earth-surface phenomena can be recorded at a distance because of the capacity of certain types of sensor to detect the result of the interaction of incident energy with the phenomena. These interactions can produce, for example, variations in force fields (such as gravity anomalies) or in patterns of reflected energy (as with reflected sunlight recorded on photographic film). Although it is valid to consider other forms of energy in the context of 'remote sensing', only electromagnetic (e-m) energy is considered here since it is by far the most relevant for most field scientists.

The use of the visible wavelengths of e-m energy for recording and mapping the earth's features by aerial photography was established long before the 1960s, and during the period since 1960 sensing techniques involving the use of other wavelengths of e-m energy have assumed greater prominence. Techniques involving the use of thermal infra-red and microwave energy, for example, are forms of remote sensing with a much shorter history of use than aerial photography. Since they have had a rather longer and, in many ways, separate history of development the techniques of mapping from aerial photography have been covered separately in Chapter 3. This chapter is devoted mainly to the treatment of those forms of remote-sensing which operate in the non-photographic parts of the e-m spectrum and to techniques utilising imagery from satellites.

Furthermore, although remote-sensing techniques used from aircraft or spacecraft have also been used extensively for imaging the atmosphere (weather satellites) and the oceans, the treatment here has been restricted to remote sensing of the land surface.

4.1.1 THE E–M SPECTRUM

The most common form of energy used in remote sensing is electromagnetic energy, and the fundamental source is the sun. It is convenient to use the 'wave model' of energy to appreciate the characteristics of e–m energy. The sun radiates a wide range of wavelengths of e–m energy into space but they all have the common characteristic of travelling with constant speed (the speed of light in vacuo, $c = 3 \times 10^8$ m. sec^{-1}). The complete range of wavelengths is referred to as the 'e–m spectrum' (Fig. 3.9).

On reaching the earth's upper atmosphere, not all wavelengths pass through to the surface. This is due to selective absorption by the atmospheric blanket. Fortunately, most of the rays which are harmful to man (such as X-rays) are removed by this mechanism.

Since even those wavelengths which are transmitted to the earth's surface are to some extent absorbed, it is important to be aware of which portions of the e–m spectrum are affected least by the absorption (the 'atmospheric windows', see Fig. 4.1).

4.1.2 ENERGY–SURFACE INTERACTION

On reaching the earth's surface, the e-m energy interacts with the surface features by being reflected off, transmitted through or absorbed by the surface to a greater or lesser extent. The degree to which each of these mechanisms takes place depends on both the wavelength of the e–m energy and the nature of the surface material. It is because the materials on the earth's surface exhibit variations in their responses to incident e–m energy that we are able, using the appropriate sensor, to create a record or image which permits multiple discrimination of these features. The changing response of an object across a range of wavelengths of e–m energy is often a crucial discriminant, and is called the 'spectral signature, (or spectral response) of the material'.

4.1.3 'REFLECTED' AND 'EMITTED' ENERGY

The portion of the e–m energy which passes into the atmosphere and is captured by a sensor (in an aircraft or spacecraft) is of greatest relevance to remote sensing, since that is the energy which will be incident on the sensor. The crucial energy segments are, therefore, the 'reflected' energy and the 'emitted' energy. The latter is composed partly of the absorbed energy which is later re-emitted and partly of the energy emitted as a result of the earth itself being a secondary source of radiant energy.

At shorter wavelengths (visible and near-infra-red), the reflected component is dominant, but in the middle- and far-infra-red and in the microwave region, emitted energy is of far greater significance. The detectors which are most sensitive to the different parts of the spectrum must be carefully selected for optimum performance.

As the wavelength of e–m energy increases, the energy intensity (and hence the ability to record a strong signal at the sensor) diminishes. In the microwave region the natural emissions are so weak that a form of sensing, known as imaging radar, has been developed, which is based on the use of microwave energy which is artificially generated in the aircraft ('active' system). This is an alternative to the 'passive' system which is based on the detection of ambient microwave emissions from the earth's surface. Sensing systems operating in the visible and infra-red parts of the spectrum are usually 'passive', although the uncommon technique of night-time photography using artificial illumination by flares would be classed as an 'active' system.

4.1.4 RESOLUTION

In the context of remote sensing there are several meanings of the word *resolution*: spatial, spectral, radiometric and temporal.

Spatial resolution
This indicates the physical size of the smallest feature or the closest separation of two features which can just be distinguished by the imaging system. Systems which operate at shorter wavelengths produce better spatial resolution than those operating at longer wavelengths. The camera–film system generally produces the best spatial resolution, other things being equal. The '*ground resolution element*' (GRE) is related to the wavelength, the diameter of the energy collector and the sensor altitude and is given by the following expression:

$$GRE = \frac{1.2 \times wavelength~(\lambda) \times altitude}{collector~diameter}$$

An aerial camera with lens diameter 10 cm at an altitude of 200 km should be able to resolve objects of about 2.4 m (taking $\lambda = 1~\mu m = 0.0001$ cm). An infra-red sensing device operating at the same altitude with the same collector diameter and $\lambda = 10~\mu m$ should resolve only about 24 m. Loss of spatial resolution is, therefore, one of the disadvantages of non-optical imaging. With a scanner system (sect. 4.2.3), the nominal ground resolution is given by the instantaneous field of view (IFOV), which is the size of the smallest ground unit which registers a separate signal. Features which are smaller than the IFOV may actually be imaged if they have a high enough contrast against the background (e.g. some roads are discernible on satellite imagery even though they may be narrower than the IFOV). The ground resolution and the IFOV, therefore, do not necessarily have the same value.

Spectral resolution
This refers to the number and width of the segments (or bands) of the e–m spectrum which are covered by a sensing system. Standard black and

white photography, for example, covers the range of visible light (400–700 nm wavelength) in one band and, therefore, has a relatively poor spectral resolution. Normal or true colour photography, however, which consists of a three-layer emulsion effectively splits the same range of wavelengths into three spectral bands: blue sensitive (400–500 nm), green sensitive (500–600 nm) and red sensitive (600–700 nm). (Refer back to Ch. 3).

Generally speaking, by increasing the spectral resolution of a system (more and narrower bands) the potential of the system to discriminate between features is improved. Features which may have rather similar reflectances over a broad band may differ in detail if the spectral interval of sensing is narrowed. The use of several bands of the spectrum in conjunction is referred to as 'multispectral' sensing (see sect. 4.5).

Radiometric resolution

This refers to the number of discrete steps which can be recorded over the operational waveband interval of a particular sensor. In the case of panchromatic photography, this refers to the number of different grey levels which make up the tonal range of the photography. With non-photographic systems, radiometric resolution refers to the number of steps or intervals on the scale of signal values.

Temporal resolution

This is indicated by the time interval between successive overpasses of the sensor when the imaging is repeated. The use of repeat coverage may be necessary when the phenomena of interest undergo significant changes with the passage of time. The changes may be significant in themselves as, for example, in monitoring the movement of potential fishing grounds via the detection of phytoplankton blooms, or the freezing and thawing of a commercially significant waterway. The changes may also form part of a pattern which when taken together over a significant time-span may improve the accuracy of identifying certain features as, for example, the identification of agricultural crops. The temporal resolution required will vary according to the application, and the user must establish what would be appropriate in each case. With airborne imaging, each repeat mission incurs more or less the same costs and so the costs of introducing temporal resolution are high. In the

case of satellite imaging, although the initial costs are high there is automatic repeat coverage due to the orbital mode of operation. The temporal resolution, or repeat period, will depend upon the orbital characteristics of the satellite.

4.2 THERMAL INFRA-RED SENSING

Any object is potentially a source of e–m energy as a result of its temperature. When the temperature is sufficient to cause the object to emit energy in the form of heat, the wavelengths emitted are predominantly in the thermal infra-red portion of the spectrum (generally taken to be between approximately 3 μm and 25 μm).
(Note: 1 μm = 10^{-6}m.)

At about 3 μm wavelength, more than half the ambient energy is due to the reflectance of incident solar energy, but with increasing wavelength the energy emitted from the surface becomes progressively the major portion of the ambient energy. At wavelengths from one micrometre upwards, photographic emulsions are no longer suitable for sensing the ambient energy from the surface, and so other forms of detector must be used.

Furthermore, due to the differential absorption characteristics of the atmosphere, only certain limited regions within the infra-red part of the spectrum (the 'windows') are suitable for thermal infra-red sensing of the earth's surface (Fig. 4.1).

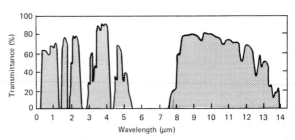

Fig. 4.1 Atmospheric transmittance of electromagnetic energy.

4.2.1 TEMPERATURE-ENERGY RELATIONSHIPS

The signal created as a result of recording the energy emitted by the earth's surface features is influenced both by the temperature of the feature and by its relative efficiency as an emitter of heat energy. The total power emitted (M) per unit area is described by the Stefan–Boltzmann law.

$$M = \sigma\epsilon\,(T)^4$$

where: M is the power emitted per unit area (in W cm^{-2});

σ is the Stefan–Boltzmann constant;

ϵ is the emissivity of the body; and

T is the absolute temperature of the body (K).

Since remote sensing is mostly concerned with comparing the power emitted by different objects or targets, it is the relative power emitted which is important, and so the value of the constant σ may be disregarded.

The emissivity ϵ of a body is a measure of its performance as a radiator compared to the so-called 'black body', which has an emissivity of 1.0.

$$\text{Emissivity } (\epsilon) = \frac{\text{Radiant emittance of an object at a given temperature}}{\text{Radiant emittance of a 'black body' at same temperature}}$$

Since a true 'black body' does not exist in reality, all objects have an emissivity of less than 1.0. However, as may be inferred from the above equation, two objects or surfaces at more or less the same temperature may record quite differently if they have significantly different emissivities (e.g. ice, $\epsilon = 0.95$, and snow, $\epsilon = 0.85$, approximately). Likewise, two objects with similar emissivity values may produce different signals if they are at different temperatures. The final appearance of a thermal infra-red image should be seen, therefore, as being due to a combination of emissivity and temperature (Fig. 4.2).

In general, the energy emitted at thermal infra-red wavelengths is largely due to the temperature of the object. However, the wavelength of peak emission is closely tied to the specific absolute temperature of the body, and if an object at a different temperature is the primary interest, then the wavelength of peak emission will alter. This relationship is described by *Wien's Displacement law* which broadly states that as the temperature of the object or source rises, then the peak emission shifts to progressively shorter wavelengths.

$$\lambda_{max} = \frac{2897}{T\,(K)}$$

Some examples of the relationship between temperature and wavelength of peak emission are given for some common energy sources in Table 4.1.

Table 4.1 Relationship between absolute temperature and wavelength of peak emission for some common energy sources

Object (at surface)	Absolute temp. (K)	Power at λ_{max} (W cm^{-2})	Range of wavelengths (μm) λ_{max}		
Sun	6000	7000	0.3	0.48	
Hot fire	1160	10	0.8	2.5	3.2
Small fire	650	1	1.4	4.5	16
Air temp. (Earth's					29
data surface)	300 K	0.04	3.2	9.7	infinite

4.2.2 DETECTORS

For remote sensing using the sun as the prime energy source, a sensing system operating at a wavelength of around 0.48 μm would be the most efficient. As it happens, this falls within the range of visible light (0.4–0.7 μm), and photographic film is therefore well suited as a detector for recording energy patterns having the sun as the prime source. If the other sources are considered then photography is not well suited to their wavelengths of peak emission (Table 4.1). For recording energy emitted by these other sources different forms of sensor are used which are more precisely responsive to the wavelengths of peak emission.

For thermal infra-red sensing the sensor which is generally used is some form of crystal detector, which is maintained at a very low temperature (Table 4.2).

A thermal infra-red sensing system designed mainly for detecting emissions from surface fires (e.g. in forest fire surveillance) would operate best in the 3–5 μm waveband, using an indium antimonide crystal detector. However, for recording

Fig. 4.2 Day-time (top) and night-time (bottom) images of thermal infra-red emissions (8–14 μm).

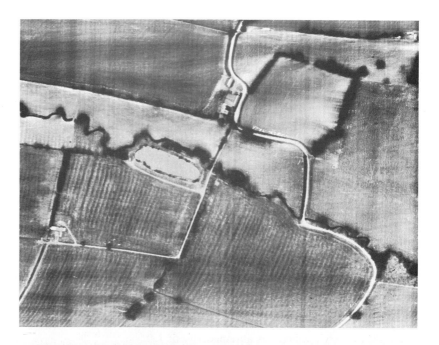

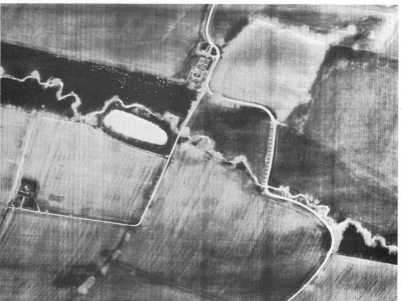

Table 4.2 Range of sensitivity of crystals commonly used in thermal infra-red imaging

Type of crystal	Spectral range	Crystal cooled to
Indium antimonide	3–5 μm	77 K (−196 °C)
Mercury-doped germanium	5–14 μm	26 K (−247 °C)
Mercury cadmium telluride	8–14 μm	77 K (−196 °C)

thermal emissions from other earth surface features (at around 300 K temperature) a crystal sensitive to longer wavelengths (mercury cadmium telluride, 8–14 μm) would be more appropriate, since the peak emissivity for most terrain features is at around 10 μm wavelength (Table 4.1).

The function of the crystal detector is to permit the recording of data about thermal emissive patterns of energy at the earth's surface by transforming the power 'caught' by the sensor at any instant into a signal which can later be analysed to allow inferences to be made about the pattern of thermal emissions from the surface. The detectors listed in Table 4.2 have a very rapid response time to the energy incident on them (less than a microsecond) and are, therefore, suitable for use in an airborne imaging system where the area of ground being sensed is changing very rapidly.

4.2.3 THERMAL INFRA-RED IMAGING

In aerial photography the ground is imaged by an area of detector elements (the photographic emulsion). By contrast, thermal infra-red imaging involves the use of a 'point' detector (the supercooled crystal). To achieve coverage of an area of ground, a scanning mirror is made to rotate across the ground track of the aircraft such that the thermal emissions from a series of small contiguous ground areas, along a line normal to the flight direction, are allowed to impinge on the crystal detector. The rapid response time of the detector ensures that each of these small ground segments will register as a separate signal on the magnetic tape record of the signals. As the aircraft moves forward the scanning mirror will trace out a series of parallel narrow swaths on the ground, which

eventually will ensure recording of the thermal emissions for an area of ground below the aircraft, normally within a viewing angle of between 90° and 120° (Fig. 4.3). The scan rate of the mirror can be adjusted according to the aircraft's altitude and speed to ensure that there are no gaps or overlaps between the swaths. Such a sensor is called a 'linescanner', and although here described for thermal infra-red sensing, the system is more widely applied as, for example, in a multispectral scanner system (MSS, see sect. 4.5.2).

At any instant, the small angular field of view will be receiving the energy emitted by a limited ground area (called the 'ground resolution element' (GRE) or instantaneous field of view (IFOV)).

The scan mirror directs the thermal radiation on to the crystal, and the temperature change induced in the crystal alters the electrical resistance of the detector. These changes in electrical resistance, from one ground resolution element to the next, are recorded in analogue form on magnetic tape. Playback of the magnetic tape can later be used to produce intensity modulations of a light spot on a cathode ray tube (CRT). In this way the differences in energy emitted from the surface can be portrayed as grey tones on a photographic type of image. This image formation can also be carried out in near real-time on board the aircraft by positioning a film recorder in front of a CRT on which the intensity modulations of the light spot are displayed, a line at a time.

Fig. 4.3 Thermal infra-red line-scanner: mode of operation.

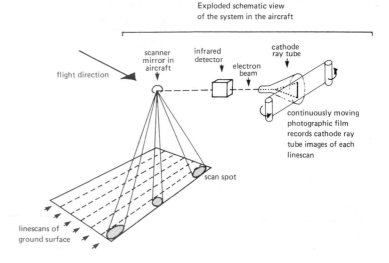

4.2.4 APPLICATIONS OF THERMAL INFRA-RED IMAGING

Wherever features can be identified on the basis that their thermal emissions differ from their surroundings, then there is potential for the use of thermal infra-red imaging. If the difference between the power emitted by a feature and its surroundings is greater than the limiting 'power resolution' of the system, then thermal imaging could be used successfully. The feasibility of using thermal linescan for a particular application can usually be established in advance by means of preliminary calculations. An example of when such calculations are necessary is where it is required to detect and enumerate warm-blooded animals as part of a census of wildlife. Where the animals are more readily counted at night, thermal linescan imaging may be the best method.

An illustration of this approach is given by the following worked example.

PROBLEM

A white-tailed deer (emissivity 0.90) in winter has a coat temperature of 260 K. Viewed from above, the deer presents an effective surface area of about 5 ft². A thermal linescanner is available with an angular IFOV of 0.003 radians and a temperature resolution of 0.5 °C. If the thermal linescanner is flown at 2000 ft altitude will the deer be discriminated against the cold ground surface (emissivity 0.92, temperature 255 K) in the resulting thermal imagery?

From a flying height of 2000 ft the size of the ground resolution element (GRE) will be given by (2000 ft $\times$ 0.003 rad)^2 or 6 ft by 6 ft (36 ft²). Since the scanner combines the signals from all objects in the GRE into a single average response, in the ground cell containing the animal the white-tailed deer will supply 5/36 of the signal, with the rest coming from the background.

The power off the ground is given by $M = \sigma\epsilon\,(T)^4$ (see sect. 4.2.1), where M is the power emitted, ϵ is the emissivity, T is the absolute temperature and σ is the Stefan–Boltzmann constant.

What is required is to compare the power given off by a ground cell containing a deer with that of

a cell consisting only of cold ground surface. The relative power difference between the two cells should then be compared with the relative power difference represented by a temperature difference of 0.5 °C (the limiting thermal resolution of the thermal linescanner).

For the cold ground surface

$$M = 0.92\,(255)^4 = 3.89 \times 10^9 \text{ units of relative power,}$$

and for the white-tailed deer

$$M = 0.90\,(260)^4 = 4.11 \times 10^9 \text{ units of relative power.}$$

(In both cases the Stefan–Boltzmann constant is omitted since we are interested in the relative power difference.)

The average power emitted by the ground cell containing deer and background is

$$M\,(\text{deer + background})$$
$$= \frac{(5 \times 4.11 \times 10)^9 + (31 \times 3.89 \times 10^9)}{36}$$
$$= 3.92 \times 10^9 \text{ units of relative power.}$$

This is (0.03×10^9) units higher than the signal from the background alone. The difference in relative power between an average background temperature of 0°F (255 K) and the same surface 0.5 K warmer is given by

$$M\,(\text{ground at } 255.5 \text{ K}) = 0.92\,(255.5)^4$$
$$= 3.92 \times 10^9 \text{ units of relative power}$$
$$\text{but } M\,(\text{ground at } 255 \text{ K}) = 3.89 \times 10^9 \text{ units of relative power,}$$

so the difference in emitted energy which is just detectable is (0.03×10^9) units of relative power.

This is the same as the difference in signal between a ground cell with a deer and one without. In order to be certain of detecting the deer the aircraft should, therefore, fly lower so that the ground cell size will be smaller and consequently the deer will occupy a larger proportion of the ground cell area.

Other specific areas where thermal infra-red imaging has been found to have advantages include: forest fire detection and surveillance; monitoring ice formation in commercially important seaways; detection of oil slicks; and tracing the

Fig. 4.4 **Night-time thermal infra-red (8–14 μm) of an urban area.**

effects of warm water discharge into the ocean from coastal power stations. In recent years, the soaring cost of all forms of energy has resulted in an increasing interest in energy conservation measures. Large industrial firms and institutions such as hospitals and universities have been concerned to establish the efficiency of their insulation measures, and consequently there has been considerably increased activity in the field of thermal imaging, since it can, under suitable controlled conditions, produce data on heat loss from buildings (Fig. 4.4).

4.3 IMAGING WITH MICROWAVE RADAR

4.3.1 GENERAL CONSIDERATIONS

In the microwave region of the e–m spectrum (from a few mm to a few tens of cm wavelength) the strength of naturally occurring emissions is very weak. Although these 'passive' microwaves can be recorded using highly sensitive radiometers, the most common imaging system in this part of the spectrum uses 'active' microwave energy in the form of a radar imaging system.

Radar (radio detection and ranging) was originally developed for detecting aircraft from the ground. Imaging radar reverses the procedure by having the radar system on the aircraft, with the artificially generated microwave energy directed towards the ground via an antenna.

Although microwaves are about 100 000 times longer than wavelengths of visible light, they are still short enough to resolve detail fine enough for most earth science applications. Microwaves also exhibit some of the surface penetration characteristics of the much longer radio wavelengths, which compensates to some extent for the perceptibly poorer spatial resolution when compared with aerial photography.

The spatial resolution of radar imagery in fact varies in both the along-track and the cross-track directions. Along-track (or azimuthal) resolution

117

depends on the radar angular beamwidth and on the range to the ground, and the cross-track (or range) resolution depends on the radar pulse length and the ground incidence angle. These complications mean that it is no simple matter to use radar imagery for standard precision mapping tasks. The most significant improvement of recent years has resulted from the development of synthetic aperture radar (SAR), whereby the azimuthal resolution is effectively made to be half the antenna length and to be independent of the range or altitude. This has facilitated the adoption of imaging radars at very great range on satellite platforms (for a fuller explanation see Ulaby *et al.* 1981).

Since an imaging radar system does not use the waveband of visible light, such a system can operate by day or night. Furthermore, microwaves are generally little affected by suspended particles in the atmosphere, such as water vapour, and so imaging radar has an ability to penetrate cloud cover, more so at the longer wavelengths. The wavelengths commonly used in imaging radar are given in Fig. 4.5. Radar is wavelength specific, in contrast to visible light or infra-red systems which generally cover a range of wavelengths (or waveband interval). Another significant difference between radar and other systems is that polarised energy is used. When the stream of transmitted microwave pulses is constrained to vibrate in a vertical plane then the energy is said to be *vertically*

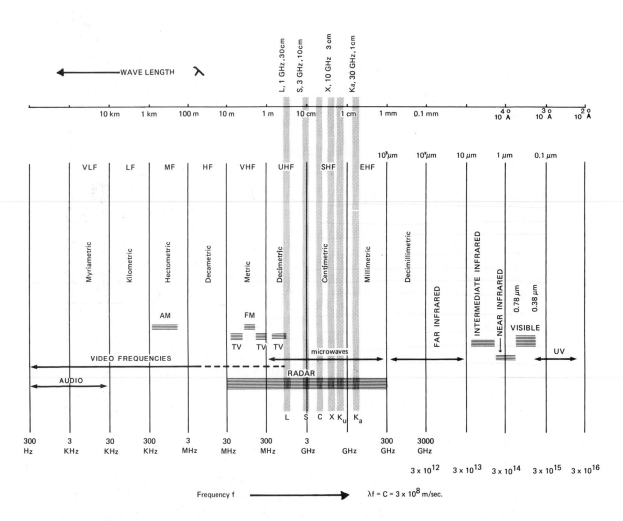

Fig. 4.5 Radar wavelengths and the electromagnetic spectrum.

Plate 1 True-colour (left) and False-colour infra-red (right) aerial photography from 500 m above ground, near Elgin (Scotland), showing characteristic water definition and superior vegetation discrimination of colour infra-red film.

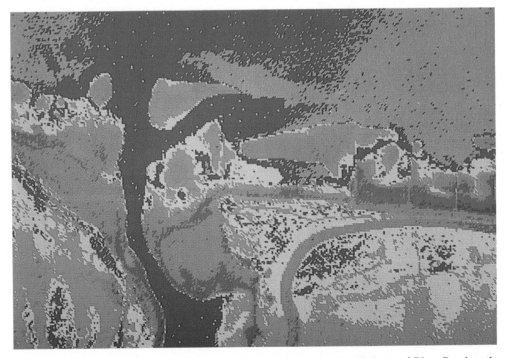

Plate 2 Night-time airborne thermal infra-red (8–14 μm) image showing discharge of River Don into the North Sea at Aberdeen.

Plate 3 Colour composite of Landsat MSS image of part of Scotland, from Inverness/Moray Firth (top centre) to Stirling/Firth of Forth (bottom right). Band 5(0.6–0.7 μm) in red and band 7 (0.8–1.1 μm) in green. Nominal spatial resolution 79 m.

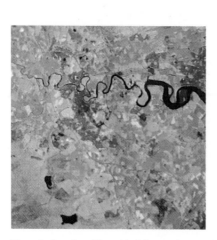

Plate 4 Landsat Thematic Mapper image of vicinity of Stirling/Alloa and River Forth. Bands 2 (0.52–0.60 μm), 5 (1.55–1.75 μm) and 4 (0.76–0.90 μm). The area (left) covers about 3 per cent of a TM scene and area (right) is a selected 3× enlargement. Nominal spatial resolution 30 m.

Plate 5 Landsat Thematic Mapper image of part of Sussex, near Arundel. Bands 2 (0.52–0.60 μm), 4 (0.76–0.90 μm) and 5 (1.55–1.75 μm).

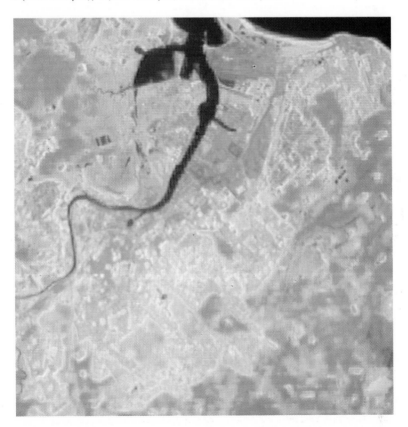

Plate 6 Middlesbrough (Cleveland) depicting levels of heat emission from an urban area. Landsat Thematic Mapper image with the thermal emissive band 6 (10.4–12.5 μm) in red, band 3 (0.63–0.69 μm) in blue and band 4 (0.76–0.90 μm) in green.

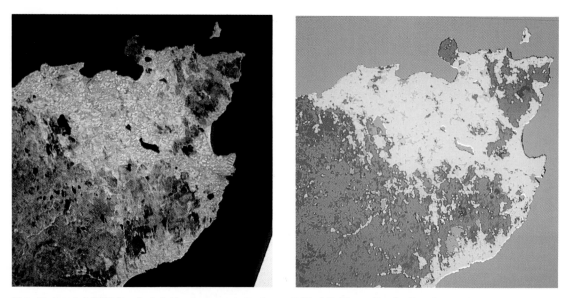

Plate 7 Landsat MSS (bands 4–5–7) colour composite image (left) of Caithness (Scotland) and part of a land cover classi-fication (right) based on digital image analysis of terrain reflectances. (KEY: yellow = cropland, purple = heathland, dark green = woodland, light green = grassland, blue = water, red = built-up).

Plate 8 A land cover classification of Tayside Region (Scotland) based on a semi-automated classification of Landsat digital MSS data. KEY as for Plate 7. (See 4.6.6.)

polarised (V). If constrained in a direction at right angles to this, then it is *horizontally polarised* (H). Furthermore, the radar apparatus can transmit and receive the same polarisation (e.g. HH, like-polarised) or on different polarisations (e.g. HV, cross-polarised). The facility to obtain like- and cross-polarised imagery offers an additional capability in discriminating surface features which offsets, to a degree, the poorer spatial resolution.

In practice, an imaging radar system operates by directing a stream of artificially generated microwave pulses towards the ground from an aircraft. The reflection of the microwave energy back to the aircraft depends on such details as feature orientation and slope, surface dielectric properties and surface roughness in relation to the wavelength. Furthermore, since the exact location of features within the radar beam is dependent on range discrimination (i.e. differences in slant distance from aircraft to ground features) it follows that the radar microwave energy should be directed to the side of the aircraft (side-looking airborne radar, SLAR) in order to avoid the range ambiguities

which exist with features close to the vertical beneath the aircraft (Fig. 4.6)

4.3.2 SURFACE ROUGHNESS AND THE INTENSITY OF ACTIVE MICROWAVE BACKSCATTER

The concept of surface roughness is dependent on the wavelength of operation, since what may appear to be 'rough' at visible wavelengths is probably 'smooth' at microwave wavelengths. Moreover, within the microwave region a rough surface at short wavelengths (e.g. X-band, $\lambda = 32$ mm) may appear smooth at longer wavelengths (e.g. L-band, $\lambda = 235$ mm). This partly accounts for the different appearance of the same surface on radar images taken at different wavelengths (Fig. 4.7).

A *rough* surface acts as a diffuse reflector, scattering incident energy in all directions, including a portion (the radar *backscatter*) which is returned in

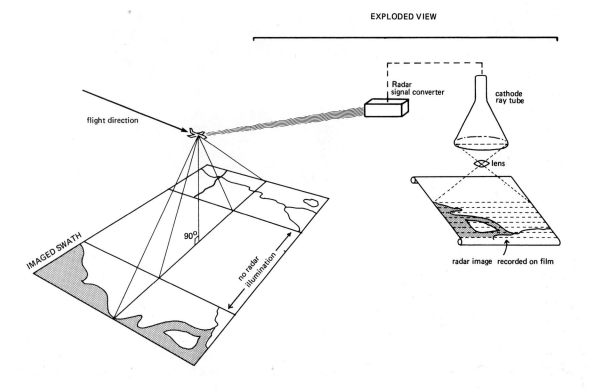

Fig. 4.6 Airborne imaging radar: mode of operation.

Fig. 4.7 Simultaneous X-band (λ = 3.2 cm, top) and L-band (λ = 23.5 cm, bottom) synthetic aperture radar imagery.

the direction of the antenna in the aircraft. It is this component which constitutes the radar return signal.

A *smooth* surface reflects the incident energy in one direction (specular reflector, like a mirror). If the smooth surface is at right angles to the incident radar beam then the energy returned to the aircraft produces a high-intensity signal. If the surface is at any other angle to the radiant beam then none of the microwave energy is returned to the aircraft. This is the case with most water surfaces and explains why smooth water almost invariably

appears black in a radar image. Where the water surface is disturbed, as for example where waves are present, then radar imagery will tend to contrast a high-intensity return from the disturbed water with a low-intensity return from the rest. This characteristic makes radar imagery highly suitable for monitoring sea surface conditions.

Fields of crops are generally diffuse reflectors at most wavelengths. Other surfaces, such as roads, car parks, airport runways and water may be diffuse reflectors in the visible region but mirror-like in the microwave region. Radar imagery, therefore, generally contains more specular reflectors than thermal infra-red or visible waveband imagery.

Sometimes, the juxtaposition of a horizontal and a vertical surface in the landscape creates a sort of dihedral reflector (e.g. between a building and a road). If the orientation of such a natural corner reflector is towards the aircraft then most of the incident energy will be backscattered directly to the antenna, and will produce a very intense signal for the junction of the two surfaces. This effect is most clearly observed on imagery of urban areas, especially where street orientation is parallel to the flight direction. Streets running at right angles to the flight direction do not display this effect. The effect is most likely to be noticed on imagery of cities with a gridiron street pattern. Since the orientation of the corner reflector in relation to the flight direction is critical for this to happen, it is referred to as the 'cardinal' effect (Fig. 4.8).

Natural corner reflectors also occur in other situations. For example, this effect may pick out the join of river surface with river bank or a bridge across a river, or the vertical crash barrier dividing a motorway. The detection and recording of a terrain feature by active microwave is therefore dependent on the favourable backscatter of the incident microwave beam to the aircraft. Visible and infra-red waveband sensors primarily record differences in the colour, composition and tempera-ture–emissivity of the surface. Radar, however, responds primarily to surface roughness and geometry and, secondly, to the composition of the material. The appearance of features on a radar image can, therefore, be quite unlike their appearance on a visible waveband photograph or on a thermal infra-red image.

The intensity of microwave returns is influenced both by the radar system properties and by terrain properties, as follows:

1. *Radar system properties* include frequency (wavelength), depression angle and polarisation.
2. *Terrain properties* include surface roughness (in relation to wavelength), dielectric constant and terrain slope.

The orientation of the ground relief in relation to the flight direction is an additional factor which touches on both radar system and terrain.

The dielectric constant is related to the composition of the material as it affects the ability to conduct the particular frequency of microwave energy. The complex interaction of radar system properties and terrain characteristics makes it extremely difficult to extract detailed information on surface composition without proper calibration of the radar returns. A great deal of systematic research into the effects of the above parameters on the resulting radar signal has been carried out, in particular at the University of Kansas. The distillation of the results of much of that research is to be found in the text by Ulaby *et al.* (1981) to which readers are referred for a detailed explanation of radar imaging systems. A useful diagram which attempts to summarise the dominant controls on radar backscatter throughout the full range of local incidence angles has been prepared by Ford *et al.* (1983, Fig. 4.9).

4.3.3 APPLICATIONS OF IMAGING RADAR

The value of imaging radar as a remote sensing system is most clearly evident in applications where other forms of imaging have been tried and failed. The most striking advantages derive mainly from its all-weather capability and the economy of coverage achieved by a side-looking mode of operation. The unusual rendition of features, resulting

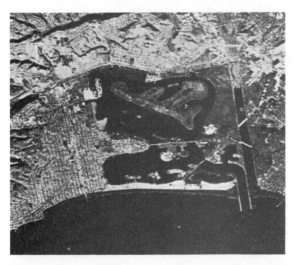

Fig. 4.8 The 'cardinal' effect on radar imagery produced by a gridiron street pattern.

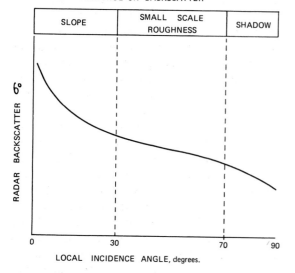

DOMINANT CONTROL ON BACKSCATTER

| SLOPE | SMALL SCALE ROUGHNESS | SHADOW |

RADAR BACKSCATTER $6°$

LOCAL INCIDENCE ANGLE, degrees.

0   30   70   90

Fig. 4.9 General relationship between radar backscatter and local incidence angle.

from their microwave backscattering characteristics, complicated by the radar system and terrain factors already mentioned, means that a proper appreciation of how the image is created is necessary for optimal interpretation of the information content.

Many terrain analysis studies based on topographic relief, slope, surface texture, primary vegetation and drainage benefit from the peculiar characteristics and distortions of radar imagery. A comprehensive study of the portrayal of drainage networks on radar imagery was carried out by McCoy (1969). Based on the interpretation of radar images at 1:200 000 scale, he concluded that certain hydrological information (total network length or total number of stream segments) extracted from the radar imagery was equivalent to the information depicted on topographic maps at 1:24 000 scale. The accurate extraction of these drainage basin parameters is attributed to the detailed depiction of the drainage lines, even down to the smallest tributary slopes. Reliable measurements of basin area and basin perimeter were also obtained from the real-aperture imagery.

Since the total network length and the drainage area are both strongly correlated to stream flow, these radar-derived measurements are potentially useful for estimates of run-off. All of this suggests

that radar imagery might be a useful tool to assist in water supply calculations for remote parts of the world.

Quantitative data relating to slope and relief can be extracted from radar imagery by making use of such features as radar layover, foreshortening and shadowing. Individual slope measurements may be derived from radar foreshortening and radar shadows can be used to produce regional slope values (see Lewis 1976).

Probably the first major project to utilise the all-weather, cloud penetrating capabilities of imaging radar was the survey of 20 000 km² of Darien province, Panama, in 1968. This used a real-aperture radar system. This project produced imagery of an area which had defied attempts to obtain aerial photography for almost twenty years due to persistent cloud cover.

The first of the extremely large-area surveys by radar were initiated by Venezuela and Brazil, in 1971, to provide a base for planning and development of rain forest areas in the Amazon basin. More than 4 000 000 km² were covered, and image mosaics produced, in less than a year. Brazil established Projecto Radam to involve specialists such as geologists, foresters and soil scientists in the interpretation of the radar imagery for a resources inventory. Among the results of this project have been the discovery of important mineral deposits, previously unknown volcanic cones and rivers and the selection of provisional routes for sections of the Trans-Amazonas Highway.

Other applications of radar imagery include the recognition of types of faults, lineaments and other tectonic features relevant to the study of continental drift. Hunting Surveys Ltd, have produced a SEASAT radar mosaic of the United Kingdom from which regional geological structure and lineaments have been interpreted (Fig. 4.10). New power stations should not be sited on or near to active geological faults and radar imagery is frequently useful for preliminary analysis of site selection.

Radar reflectances cover a very wide range of brightness values, which usually exceeds the tonal range of black and white film. Consequently, radar images are often better represented by colour coding the brightness levels by utilising a computer-linked colour display system, thus going directly from the digital value of the radar signal to a colour

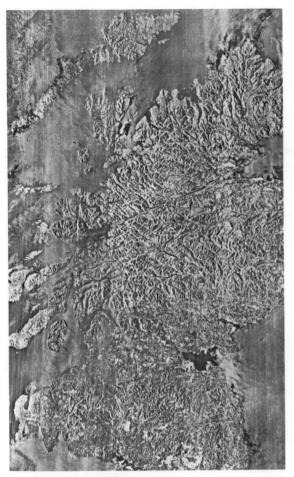

Fig. 4.10 Part of a Seasat radar mosaic of the UK, showing Scotland.

representation on a TV monitor. Moreover, with the greatly improved spatial resolution now attainable with synthetic aperture radar, composite images can be produced from satellite data combining the spatial resolution of Seasat SAR (25 m) with the better spectral resolution of Landsat (four spectral bands in visible and near-infra-red wavelengths). This combination can produce an improved accuracy in, for example, crop classification when compared with the results using either sensor type on its own (Nüesch 1982).

4.3.4 IMAGING RADAR IN THE 1980s

Just as in the 1970s earth imaging from satellites

meant essentially using the Landsat series, the 1980s is already shaping up to be the age of satellite imaging radar. This is largely due to the greatly improved ground resolution attainable by synthetic aperture radar and the growth in the capacity of modern digital computers to handle the enormous increase in data flow from an orbital radar system compared with the Landsat multi-spectral scanner system. The value of a satellite radar for imaging cloud-prone areas of the earth was demonstrated by Seasat in 1978. The second test flight of the Space Shuttle *Columbia*, in 1981, included a Shuttle Imaging Radar (SIR-A) experiment which obtained imagery of 10 000 000 km² of the earth's surface (Ford *et al.* 1983). The specification for SIR-A was based largely on experience with the earlier Seasat imaging radar. Although the same wavelength was used (23.5 cm, L band), some of the parameters are quite different, resulting in a nominal ground resolution of 40 m, rather than the 25 m of Seasat (Table 4.3). Further imaging radar experiments are planned for future Space Shuttle and European Space Agency missions.

Table 4.3 Selected system characteristics of SIR-A and Seasat synthetic aperture radar. (Source: Ford *et al* 1983).

Parameter	SIR-A	Seasat SAR
Orbit		
Altitude (km)	259	795
Inclination (deg)	38	108
Radar		
Frequency (GHz)	1.278	1.275
Wavelength (cm)	23.5	23.5
Transmit pulse duration (μs)	30.4	33.4
Polarization	Like HH	Like HH
Antenna		
Dimension (m)	9.4 × 2.16	10.74 × 2.16
Look angle (deg)	50 ± 3	23 ± 3
Swath width (km)	50	100
Resolution (m)	40 × 40	25 × 25
Data recording	Onboard: optical	Ground station: digital

In areas subject to persistent haze and cloud, or where there is a complex pattern of water features, satellite radar images have demonstrated a considerable advantage over Landsat imagery (Fig. 4.11).

Fig. 4.11   Comparison of Landsat MSS (top) and SIR-A radar imagery of part of the Mississippi delta.

4.4 SPECTRAL SIGNATURE (OR SPECTRAL RESPONSE)

Remote-sensing techniques are concerned with recording the interaction of electromagnetic radiation and the earth's surface features, with a view to discriminating those features and/or ascertaining something about the condition of the features. Since the e–m spectrum covers a wide range of wavelengths and since most surface materials do not react in the same way at all wavelengths, it follows that some parts of the spectrum are more suitable than others for the discrimination of particular features.

The reflectance and emittance behaviour of a feature at different e–m wavelengths is called the 'spectral signature', or spectral response, of the feature. At visible and near-infra-red wavelengths this can be represented by a spectral reflectance curve (Fig 4.12).

Knowledge of the spectral signatures of features of interest permits identification of the part(s) of the spectrum in which the clearest separation of the features may be achieved.

4.5 MULTISPECTRAL SENSING

4.5.1 GENERAL

It may happen that the remote sensing method being used (say monochromatic photography) may not pick out the finer variations in spectral signature because the waveband of operation may be too broad (400–700 nm for black and white photography). In that case, such photography would record the average reflectance of the features over the range of sensitivity of the emulsion. However, if the components of blue, green and red reflectance could be recorded separately then it should be possible to achieve greater discrimination between, for example, types of vegetation. Recording in several narrow bands instead of one broad spectral band is the basis of *multispectral sensing*. The simplest example of this is colour photography, where there are three layers of emulsion, each sensitive to a different portion of the e–m spectrum (blue/green/red for *true* colour and green/red/reflective infra-red for colour *infra-red* photography).

Multispectral sensing is also possible with a single layer black and white photographic emulsion. The part of the spectrum exposing the film may be restricted by using an optical filter in front of the lens. In order to split the spectrum into several narrow bands requires a separate camera/film/filter combination for each narrow band, so that a multi-camera arrangement may be necessary. An alternative is to have a special camera which uses one roll of film but has several lenses, each with a different filter, to allow simultaneous recording of several narrow bands of the spectrum. The ultimate development of this multiband camera approach was the ITEK 9-lens camera, which exposed nine small images of narrow bands of the visible and reflective infra-red on to black and white infra-red film. Practical difficulties in making visual comparisons of 9 images led to the demise of this 9 lens system.

Fig. 4.12 Spectral reflectance curves of some common materials.

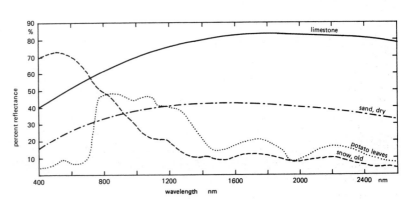

4.5.2 MULTISPECTRAL SCANNER

One of the principal difficulties with a multiband photographic system is that the detection of changes in reflectance of different features is established by visual correlation of the different images. This may be tolerable for three bands but is difficult to carry out efficiently for more than three bands.

If the reflectances in the different narrow spectral bands existed in digital form, rather than as grey tones, it would be possible to program a digital computer to make multiband comparisons much faster and more effectively than is possible by visual comparison. This is the basis of the *multispectral scanner*.

Instead of using a lens system to transmit the e–m energy on to a light sensitive surface (the film), the multispectral scanner reflects the e–m energy from the terrain on to a series of narrow-band detectors, each sensitive to a limited part of the spectrum. Variations in the amount of e–m energy falling on each detector causes measurable changes in the electrical properties of the detector. These changes (which are proportional to differences in the terrain reflectances) are recorded as analogue signals on a tape recorder (Fig. 4.13).

The analogue signal tape can later be played back through an analogue-to-digital converter and

Fig. 4.13 Airborne multispectral scanner: mode of operation.

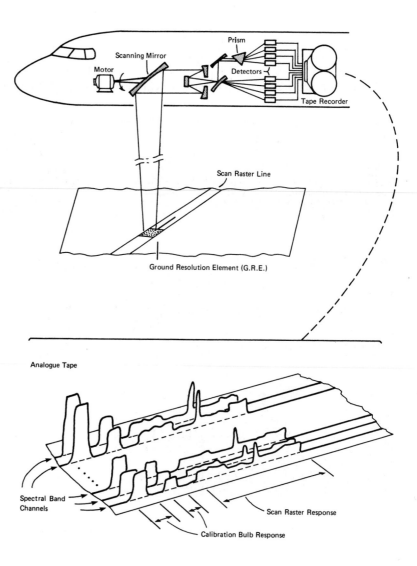

the terrain reflectance data for each ground element can be obtained in the form of a computer compatible tape (CCT). The scanning mode is similar to that already described for the thermal infra-red linescanner (sect. 4.2.3). The size of the smallest area creating a separate signal (the ground resolution element, GRE) depends on the angular instantaneous field of view (IFOV) of the scanner and the flying height above ground of the aircraft.

In theory, the number of spectral bands which can be recorded by this method is limited only by the number of separate detectors available and the ability to record a large number of channels. When the data are in digital form, a high-speed computer can carry out any number of statistical operations on the data in a minimal time period.

In practice, the maximum development of the multispectral scanner has been the 24 channel system (built by the University of Michigan). The expense of recording, processing and analysing so much data has usually resulted in only the optimal number of channels being used in any particular project, since many adjacent channels produce data which is highly correlated. In practice, often only between 5 and 12 of the available channels may be used (Fig. 4.14).

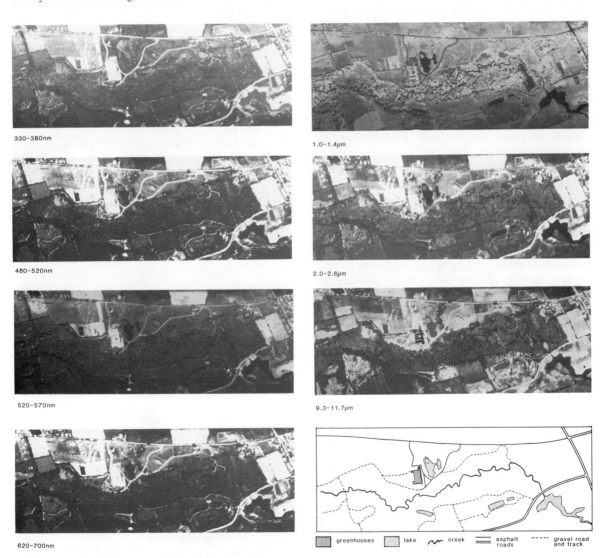

330–380nm

480–520nm

520–570nm

620–700nm

1.0–1.4µm

2.0–2.6µm

9.3–11.7µm

greenhouses   lake   creek   asphalt roads   gravel road and track

Fig. 4.14 Sample of 7 bands from airborne multispectral scanner system.

4.6 IMAGING THE EARTH'S SURFACE FROM SATELLITES

4.6.1 GENERAL

When considering earth imaging from satellites, the same spectral bands and sensors may be used as for airborne imaging. There are significant differences, however, which justify a separate treatment for earth remote sensing from satellites.

The difference in platform altitude, for example, is frequently tenfold or greater, which results in an image scale so much smaller than anything obtainable from aircraft that the normal 'rules and guidelines' for air photographic interpretation no longer apply.

The Landsat series of satellites had an altitude of 915 km and produced images with a ground scene coverage of 185 km by 185 km (or 34 225 km²). For coverage of an equivalent ground area using a civilian survey aircraft, the limiting photograph scale of around 1:100 000 would require more than 80 photographs (excluding stereo cover). There is, therefore, a significant difference in the total quantity of data to be handled and analysed for the same ground area coverage. This suggests that the synoptic coverage of space imagery might allow a different set of problems or topics to be addressed on a national, continental or global scale which are incapable of being tackled utilising aerial photography. With appropriate selection of orbit it is possible to view and image the whole earth, and to go on repeating this coverage for as long as the satellite stays up. The potential for global coverage and a long-term regular repeat cycle at minimal extra cost are additional features of a satellite system which would, in practice, be unattainable by an airborne system.

4.6.2 LANDSAT

Although the USA and the USSR have been active in space flight, both manned and unmanned, since the late 1950s, systematic imaging of the earth's surface for purposes of mapping and monitoring surface features dates mainly from the early 1970s.

Systematic monitoring of the earth's atmosphere from satellites, largely to assist in weather forecasting, dates from about 1960.

Earth imagery acquired by American satellite programmes is more freely available to other countries than are the image products of Soviet missions. Consequently, the space imagery considered here is from the US programmes.

During the 1960s imagery of the earth's surface

Orbital parameters of Landsat:

Inclination (deg)	99
period (min)	103
time of descending node (local time of Equatorial crossing)	9.30 a.m. (approx)
altitude (km)	915
repeat period (days)	18

Sensor characteristics:

Multispectral scanner Subsystem (MSS)		Thematic Mapper (TM) Landsat 4 on
micrometres		micrometres
0.5 – 0.6	BAND 1	0.45 – 0.52
0.6 – 0.7	BAND 2	0.52 – 0.60
0.7 – 0.8	BAND 3	0.63 – 0.69
0.8 – 1.1.	BAND 4	0.76 – 0.90
	BAND 5	1.55 – 1.75
	BAND 6	10.40 – 12.50
	BAND 7	2.08 – 2.35
80m (bands 1-4)		30 m (bands 1–5, 7)
		120 m (band 6)

(**NOTE:** on Landsats 1, 2 and 3, MSS bands were labelled 4, 5, 6, 7 with 1, 2, 3 used for the return beam vidicon subsystem)

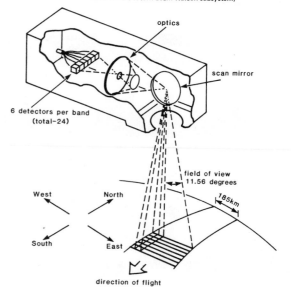

Fig. 4.15(a)   The Landsat multispectral scanner subsystem (MSS).

was acquired in a rather piecemeal fashion, on an opportunity basis, by astronauts using camera/film sensors. There was no systematic programme of coverage since earth imaging was not the primary concern of these missions. This applies to most of the photography acquired during the Gemini and Apollo missions.

It was not until 1972 that an experimental satellite with earth imaging as a primary objective was launched. Called the Earth Resources Technology Satellite (ERTS-1), later renamed Landsat, this satellite was the first of an intended series of four such experimental satellites to be launched by the National Aeronautics and Space Administration (NASA) during the period 1972–82. In an extension of the original programme Landsat 5 was launched in March, 1984. The most important

parameters of these satellites are given in Fig. 4.15.

The principal sensor on the Landsat experiments was a multi-spectral scanner system (MSS). Although it was intended that a return beam vidicon (RBV) camera system should operate alongside the MSS, due to mechanical failure this happened to only a limited extent, on Landsats 2 and 3 only. By far the bulk of the Landsat imagery is, therefore, from the MSS. Similarly the thermal infra-red band on Landsat 3 operated for only a short period. A new scanner system, the Thematic Mapper, with improved spectral and spatial resolution, was introduced on Landsat 4.

Radiation reflected from the terrain features in the four bands of the MSS is directed by the oscillating mirror on to appropriate detectors, and the changes induced in the detectors by the incident

Fig. 4.15(b) Landsat 4 and 5 index map for Britain and Ireland.

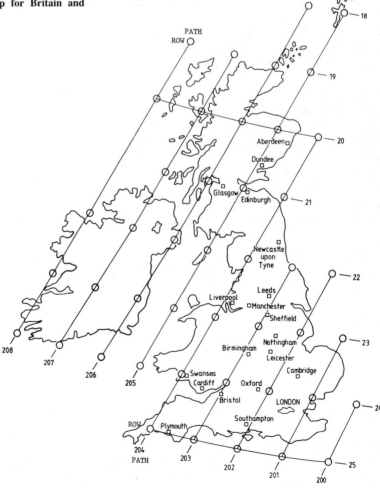

radiation are recorded as analogue signals. These signals may be transmitted, if within range, to a ground receiving station in near real-time or, alternatively, the signals may be stored on an on-board tape recorder for later play-back and trans-mission to a ground station. Since the system is based on telemetry, with no build-up and permanent storage of signals on board the satellite, the system can go on recording and transmitting signals for as long as the hardware functions. The linescanner scans 6 lines simultaneously, so that if four spectral bands are recorded then a total of twenty-four detectors are required (6 matched detectors for each band). At any instant a ground area (resolution element) of about 79 m square is the source of a separate signal for each spectral band. For each Landsat ground scene (made arbitrarily 185 km along-track, to produce a square ground area) there are approximately 2300 picture elements (pixels) in the along-track direction and 3300 pixels in the across-track direction (due to 23 m overlap of adjacent pixels, making them effec-tively 56 m wide). This produces approximately 7.6 million separate bits of signal data for each band, or more than 30 million for all four bands of one scene. These data are transmitted and recorded on magnetic tape at a ground station where the signal values are converted to digital values on a computer compatible tape (CCT). The digital data are then amenable to processing and analysis using digital computers. The signal data are also used to generate a master photographic negative for each band, from which other optical products may be generated. The digital data on CCT are theoretically on a scale of values from 0 to 255, but this is of necessity compressed to fewer than 20 grey levels on a photographic product due to the limited dynamic range of photographic film compared with magnetic tape. The photographic product, although cheaper and easier to handle and view, contains only a fraction of the detail which is present on the CCT of the same scene.

Where the fine detail is more important than the detection of large-scale features or regional trends, the digital data are of much greater value. The prime difficulty, however, is that a computer with enor-mous storage capacity and fast computing capability is required to handle such a vast amount of data. This problem will increase with the availability of data of improved ground resolution and larger number of spectral bands (such as the Thematic Mapper on Landsats 4 and 5, with 7 spectral bands and 30 m nominal ground resolution). Until fairly recently, the digital analysis of Landsat data has required the use of an expensive mainframe computer, or at least a minicomputer. Modern developments in relatively much cheaper micro-computers and the availability of data for sub-scenes on micro-floppy disks has opened up the field of digital analysis of Landsat data to many users at the low-budget end of the market (sect. 4.6.3.2).

4.6.2.1 LANDSAT DATA PRODUCTS

In the early 1970s, all Landsat data products were distributed by the US Geological Survey through the EROS Data Centre at Sioux Falls, South Dakota. With the development of ground receiving stations outside North America, a number of regional centres emerged which now act as the focus for data recording, processing and distri-bution (Fig. 4.16). Within Europe, for example, there are two Landsat receiving stations (at Kiruna in Sweden and Fucino in Italy), and satellite data distributed to each member country via a struc-ture called Earthnet. Within each country there is a National Point of Contact (NPOC) with Earthnet from which users can obtain data products. For example, the NPOC for the United Kingdom is at the National Remote Sensing Centre, Royal Aircraft Establishment, Farnborough (See Appendix 2).

The two principal forms of Landsat data prod-ucts are optical and digital. The digital product is in the form of a computer compatible tape (CCT) on which the reflectance of each ground resolution element (approximately 79 m square for Landsats 1, 2 and 3) is recorded as a number between 0 and 255. The fullest possible detail of the original signal values is, therefore, retained on the CCT. With the increasing use of microcomputers, the EROS Data Centre introduced, in 1983, Landsat digital data on 8-inch floppy disks which can accommodate all four MSS bands for a ground area of approximately 300 km^2.

Optical products are derived from the digital products, since a master film negative is generated from the original signal tape. The master negative

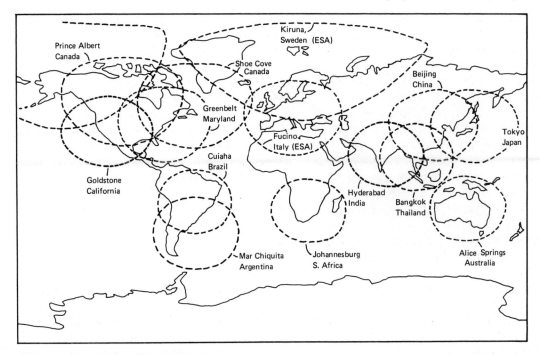

Fig. 4.16 Landsat ground receiving stations.

is used to generate second- and third-generation film positives and negatives. Photographic film cannot retain as much of the original detail as a magnetic tape and, furthermore, with each photographic stage there is an inevitable loss of some fine detail, unlike CCTs where an exact copy of the original is possible. The optical products, however, allow the production of a photographic image of a scene relatively cheaply and which can also be viewed, analysed and interpreted using fairly inexpensive equipment. The digital product requires access to a computer and suitable software for image analysis.

4.6.3 LANDSAT IMAGE PROCESSING AND ANALYSIS

Processing and analysis of Landsat data can be considered under the two types of data product: optical and digital.

4.6.3.1 ANALYSING OPTICAL DATA

Film products are available as positives or negatives (from which enlarged prints can be made) and at a scale of approximately 1:3.4 million (70 mm film chip) or 1:1 million. Each of the 4 MSS bands is in the form of a black and white image, approximately on a 16-step grey scale. Each individual image can be viewed optically, one possible arrangement being to use a 1:1 million film positive on a Zoom Transferscope type of instrument to match with common details on a larger-scale (1:250 000) map (Fig. 3.34). Revision of existing detail or the addition of specialised distributions can then be added to the base map.

Viewing and mapping from the 70 mm black and white film chips is somewhat difficult due to their small size. It has, however, been found that a microfiche viewer (of the type in widespread use in libraries) makes a highly suitable projection device for 70 mm film positives. The fixed enlargement image (usually 20 or 40 times) can then be interpreted and mapped by placing a tracing sheet over the screen (Fig. 4.17). The scale of the traced map can then be standardised to the required scale using an optical pantograph type of instrument (such as a Zoom Transferscope, Map-o-Graph or Plan Variograph). The 70 mm film chips can also be magnified by using a 70 mm slide projector, but

Fig. 4.17 **Microfiche viewer for analysis of optical products.**

these are generally less common, and the projected image is more awkward for mapping from than is the standard microfiche viewer.

Since each of the four MSS bands of Landsat records the radiation reflected in four different parts of the e–m spectrum, it is possible to produce a colour composite image which combines the attributes of all four bands. In practice, since colour formation is essentially a three-band process, only three of the four bands are used in the production of colour composites. Features which may not be separable on a single band may become discrete when viewing a combination of three bands. Moreover, the use of colour rather than grey tones improves the interpretability of the image. A colour composite of a Landsat scene can be purchased as an optical product, but the cost is considerably higher than for a single-band black and white image.

Perhaps the simplest and cheapest method of generating a colour composite image is by using diazo colour film (Malan 1981). In order to produce a colour composite image, 1:1 million scale film positives of each of three bands of Landsat MSS (usually green, red and one of the reflective infra-red bands) are in turn placed in contact with a sheet of diazo colour film and exposed to ultra-violet radiation. Sheets of yellow, magenta and cyan colour film are used for the three bands. The Landsat transparency and the diazo film are placed

emulsion to emulsion and clamped between a stiff board and a sheet of glass. This assemblage can then be exposed to direct sunlight for a few minutes. Normally a narrow test strip of film would be exposed first to get the exposure correct. It is usual, however, to make the method independent of weather conditions by using an artificial source of ultra-violet radiation, as in an automated ammonia copying machine or in a build-it-yourself flat-bed exposure box, using actinic fluorescent lamps as the u–v source. Development of the diazo film is carried out by exposing it to moist ammonia vapour in a closed container. By superimposing each of the three colour films in perfect register, a false-colour composite image of the Landsat scene can be viewed over a light table. The composite is in false-colours since the third band is from invisible reflected infra-red radiation rather than from the blue visible radiation.

Film positives of the 70 mm format may be reprojected through different coloured filters to produce a false-colour composite of a scene. Although ordinary 70 mm slide projectors could be used, this invariably runs into difficulties in achieving satisfactory registration of the three images. Because of this difficulty a number of companies have manufactured special colour additive viewers, in which four optically matched projectors are assembled and positioned on accurately aligned guide rails so that the projected images (brought to a horizontal screen using mirrors) can achieve perfect registration with a minimal amount of adjustment (Fig. 4.18). Generally, only three of the four projectors are used at any one time. The user views the projected false-colour composite image at about 1:1 million scale and can proceed to mark interpretations on to a transparent overlay on the screen. Alternatively, the user may experiment by using the controls on the instrument to alter the balance of illumination (and hence the colour emphasis) between the three bands, or to change the filter in each projector by rotating the filter wheels, which contain four filters for each projector. The results of such experimentation can be recorded on to colour slide film by mounting a camera over the screen. It is also possible to produce a contact colour print directly off the screen using the positive-to-positive colour printing process. Colour slides of the screen can be studied elsewhere using a normal slide projector or

Fig. 4.18 Colour additive viewer.

a microfiche reader, although it must be realised that some of the subtle colour variations visible on the screen may be lost in going through a further photographic stage.

4.6.3.2 ANALYSING DIGITAL DATA

Computer compatible tapes of Landsat MSS data contain more than 30 million bits of data for each 185 km by 185 km ground scene. The analysis of such data requires the use of a high-speed digital computer with adequate capacity to handle such a quantity of data.

If a general purpose mainframe computer is used then suitable software must be written to enable the user to output the data in image format and to attempt classification of the digital data. If the user has to wait until a grey scale lineprinter map of a classification is output before introducing modifications to the classification, then the whole procedure becomes rather laborious. The procedure is greatly improved if the computer has a separate imaging terminal (preferably in colour) where the result of a classification can be displayed (and photographed) and amendments made before deciding the final classification.

Some manufacturers have developed separate stand-alone image analysis systems with their own dedicated minicomputer. Certain standard preprocessing and classification procedures are hard-wired

into the system and can be called upon more or less at the touch of a button. Such a system permits manipulation of the data such as by a contrast stretch, to make the dark parts appear lighter, for example, before applying classification procedure. Classification may be based on a single band, whereby the range of reflectance values may be divided into statistically separate groups and each allocated a colour (density slicing). In addition, all four bands may be used in combination to permit a multi-spectral classification, in which features may be recognised according to a four-dimensional signature. Examples are the systems developed for the United Kingdom National Remote Sensing Centre: Plessey IDP 3000 and the GEMS image processing system (Fig. 4.19). Some of the standard algorithms available on such a system are detailed in Table 4.4.

Fig. 4.19 Digital image processing workstation.

As the nominal spatial resolution of satellite data improves from Landsats 1, 2 and 3 (IFOV of 79 m) to the Landsat 4 Thematic Mapper (IFOV of 30 m), the digital data required per unit area is increased enormously. When this fact is considered along with the trend towards charging real costs for satellite data it can be appreciated why satellite data are now becoming available for subscenes, and in the form of floppy disks for use with low-cost microcomputer-based systems.

Because microcomputer-based image analysis systems are likely to cost less than a tenth of a minicomputer-based system, there is every likelihood that an image analysis facility will increasingly become common workstations in users' labora-

Table 4.4 A selection of standard algorithms available on a GEMS or IDP 3000 image processing system

Algorithm	Use
Contrast Stretching	To modify pixel intensity values in order to obtain the best imagery for visual display.
Destriping	To eliminate striping in the image caused by imperfect matching of radiometric characteristics of the 6 detectors involved in Landsat's 6 simultaneous scan lines.
Principal Components Analysis (P.C.A.)	There is a high correlation between the 4 spectral bands of Landsat MSS. P.C.A. involves selection of a new set of orthogonal (or principal) axes such that the newly formed spectral bands are uncorrelated. This can be used to improve discrimination of similar types of ground cover.
Density slicing	Combining and coding pixels with similar intensity levels. It is useful in forming masks to mark boundaries between regions.
Geometric transformation	Ground control points (GCPs), which can be readily identified on image and map, are used to determine the transformation between image and map coordinates. This can then be used to fit the image to the map coordinate system.
Multitemporal analysis	After geometric transformation to a common projection, scenes from different times of year (or from different years) can be compared for change detection.
Digital mosaics	Involves the patchworking of images from the same sensor to form mosaics covering large areas. Poses radiometric and geometric problems since the selected scenes may have been imaged at different seasons of the year.
Image classification	To permit separation of the image into various classes, e.g. of ground cover. Makes use of interactive display system and back-up software (e.g. 'box' and 'maximum likelihood' classifiers) to enable supervised classification (i.e. using foreknowledge of the actual landcover of certain areas to permit classification of all other similar areas).

(*Source:* UK National Remote Sensing Centre, RAE, Farnborough).

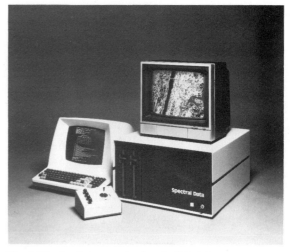

Fig. 4.20 A microcomputer-based digital image processing system.

tories. Instead of travelling to a central facility and buying time on an image processing system, the user in the late 1980s is likely to have a microcomputer-based system nearby which can cope with all but the most demanding tasks in Landsat digital image analysis. An example of this new generation of equipment was the first commercially available system (The Remote Image Processing system of Spectral Data Corporation), which was produced to the specification of the EROS Data Centre for working with Landsat digital data on 8-inch Floppy Disks (Fig. 4.20).

4.6.4 SKYLAB

Over the period May 1973 to February 1984, three Skylab manned missions obtained imagery of the earth from a variety of sensors. The theoretical coverage of the imagery was restricted to 50° north and south of the equator, because of the inclination of the orbit. At an altitude of 435 km the orbital period was 93 minutes, with a repeat cycle of 5 days. For various reasons (cost, adverse weather and demands on astronauts' time for other experiments), imagery was not obtained on a continuous basis. In general, instrument time was devoted to gathering data for specific pre-planned investigations, many of which were coordinated with aircraft and field observations at ground sites.

Table 4.5 Skylab imaging sensors

Sensor	Description	Bands (μm)	Nominal ground resolution (m)
Earth Terrain Camera	127 mm wide film. camera f/4 lens, focal length 457 mm, ground coverage 109 km on each side. 60 per cent stereo-overlap. Nominal image scale 1:950 000.	0.4–0.7 (high-resolution colour)	21
		0.5–0.7 (hi-definition black and white)	17
		0.5–0.88 (high-resolution infra-red colour)	23
Multispectral photographic camera	70 mm wide film. 6 high-precision cameras with matched optical systems. Focal length 152 mm. Ground coverage 163 km square. 60 per cent stereo-overlap. Nominal image scale 1:2 850 000.	0.7–0.8 (monochrome infra-red)	79
		0.8–0.9 (monochrome infra-red)	79
		0.5–0.8 (colour infra-red)	79
		0.4–0.7 (high-resolution colour)	46
		0.6–0.7 (Panatomic-X black and white)	38
		0.5–0.6 (Panatomic-X black and white)	46
Multispectral scanner	Conical line scan with IFOV 0.182 millirads (79 m square on ground). Swath width 74 km. Data in form of computer-compatible tapes, selected data processed to produce 127 mm film images from radiometrically corrected data.	1. 0.41– 0.45 2. 0.44– 0.52 3. 0.49– 0.56 4. 0.53– 0.61 5. 0.59– 0.67 6. 0.64– 0.76 7. 0.75– 0.90 8. 0.90– 1.08 9. 1.00– 1.24 10. 1.10– 1.35 11. 1.48– 1.85 12. 2.00– 2.43 13. 10.20–12.50	79 m for optical wave bands

(*Source:* Skylab Earth Resources Data Catalogue, NASA, 1974)

Ground coverage of the Skylab imagery is mainly confined to North and South America, with minor concentrations in Southern Europe and Japan. Details of the imaging sensors are given in Table 4.5. Most research interest in the field of mapping has been in the Earth Terrain Camera (E.T.C.) photography, since it provided the best spatial resolution of any civilian satellite sensor during the 1970s. It also provided stereoscopic coverage along the orbital track, thus permitting tests on the heighting accuracy to be performed, although it should be stressed that the E.T.C. was not a metric camera in the terms of the photogrammetrist (see sect. 3.3).

4.6.5 MAPPING FROM SATELLITE DATA

When considering the role of satellite data in mapping the earth's surface, it is necessary to distinguish between *thematic* and *topographic* maps.

Thematic maps generally depict specialised distributions of phenomena, such as vegetation or land cover types, and accuracy requirements are more concerned with the correct identification of the content (semantic information) than with the precise location (metric accuracy) of the lines giving the planimetric outline of the distributions. With topographic maps, however, the precise spatial location of features is important (both plan and height), in addition to correctly identifying the feature. The demands which topographic mapping places on the data source are, therefore, much higher than with certain thematic maps. From satellite altitudes it is more difficult, if not currently impossible, to achieve the heighting precision required for the depiction of contours on topographic maps at 1:250 000 scale and larger.

Many workers who have investigated the potential of satellite imagery, particularly of Skylab and Landsat, for topographic mapping have concluded that the level of detail in Landsat MSS and RBV imagery may be generally adequate for the plan content of 1:250 000 and smaller scales, although some elements such as roads are not always detectable where there is poor target-to-background contrast (Fig. 4.21). Photography from the Skylab Earth Terrain Camera, with its vastly superior spatial resolution (down to 17 m) is adequate for mapping planimetry at scales of 1:100 000, or 1:50 000 in some cases. Heighting accuracy requirements are much more difficult to attain, and it is thought that contour intervals closer than 50 m will be difficult to attain from satellite photography. Such a contour interval is normally associated with maps of 1:100 000 scale and smaller (Welch and Marko 1981).

Thematic mapping is not so constrained as topographic mapping by the rigid requirements for

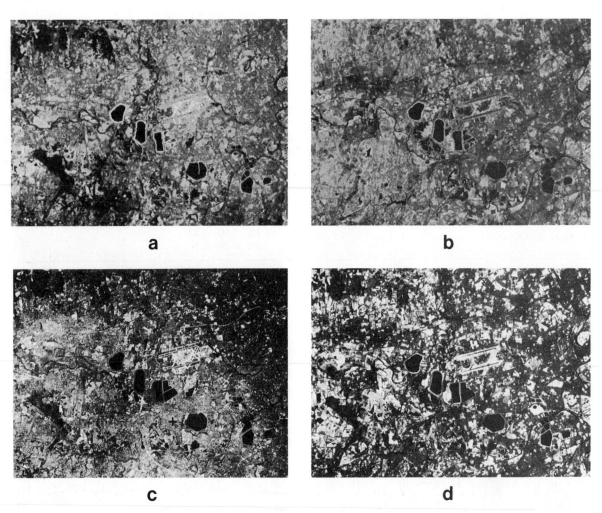

a

b

c

d

Fig. 4.21 **Comparison of image quality and resolution: (a) Landsat MSS5; (b) Landsat MSS7; (c) Landsat RBV; and (d) Landsat TM.**

geometric accuracy. The accuracy of the image content is affected by the spatial resolution (the better the resolution, the finer the detail depicted), but it is also affected by the radiometric resolution (number of grey levels) and by the temporal resolution (especially for themes, such as vegetation, which may change seasonally).

One of the more common forms of thematic map prepared from satellite imagery is a land cover types map. Since the data recorded is generally the reflectance of the surface material, the land cover types may be discriminated by their spectral reflectance characteristics. The categories in the classification must therefore be capable of discrimination on this basis. Many of the existing classification schemes are based on land use or activity, which can usually only be ascertained by ground survey. The need for a classification which is compatible with data derived from remote sensing has long been recognised. Much of the pioneering work in this field was directed by the late James Anderson, chief geographer of the US Geological Survey (Anderson *et al.* 1976). Arising out of this work, some of the significant points that should be satisfied by a classification based on remotely sensed data are:

1. 85 per cent overall accuracy in identification of land use and land cover categories;
2. the interpretation accuracy for different categories should be about equal;
3. repeatable results should be obtained from one interpreter to another and from one sensing time to another;
4. the classification should permit vegetation and other types of land cover to be used as surrogates for activity; and
5. The classification should be suitable for use with remote sensor data obtained at different times of year.

On a more general level, there is a fair measure of agreement (Ryerson and Gierman 1975; Anderson *et al.* 1976; Rhind and Hudson 1980) that a good land cover/land use classification should be:

1. hierarchical in structure, to cope with surveys carried out to different levels of detail;
2. structured to accommodate data produced from different survey technologies (ground survey, aerial photography, satellite imagery);
3. exhaustive, so that all features in the survey area are classifiable;
4. selected so that classes are mutually exclusive (i.e. any feature can fall into only one class);
5. based on what is observed, and the field work or image interpretation should be done at the smallest unit which can be differentiated;
6. amenable to computer processing in order to speed up area computations, comparisons with other geodata files and generally to expedite analysis of the data; and
7. compatible with existing land use classification schemes as far as is possible.

The classification scheme which has achieved most widespread use is that developed by Anderson and others at the USGS. (Table 4.6), often with alteration in detail in order to cope with local peculiarities in land cover.

This classification was the basis of a programme to produce by the late 1980s land cover/land use maps and statistics for the USA, utilising remote sensor data. Experience so far suggests that Landsat's 79 m nominal ground resolution data is only reliable for level 1 categories, and that data for the more detailed level 2 and 3 categories are being provided by aerial photography. In this context, the National High Altitude Photography (NHAP) programme is of particular significance since the original aim was to image, by 1984, the complete area of the co-terminous USA with 1:80 000 scale colour infra-red and black and white panchromatic stereoscopic photography, and to repeat coverage on a cyclic basis. Until such times as satellite imagery can be produced with spatial resolution of 10 m or better, then most mapping tasks in the USA which utilise remote sensing are likely to rely heavily on the high altitude photography programme, either for basic mapping or for verification. Such photography can give an effective ground resolution of less than 5 m, a value which is inconceivable using current state-of-the-art technology in non-photographic imaging from space. Most other countries, however, do not have a comparable high altitude national aerial photography programme, and so for most parts of the world regular repeat coverage of large areas will continue to depend on imaging satellite systems.

Table 4.6 The US Geological Survey Land Use/Land Cover Classification System for use with remote sensor data (abbreviated)

Level 1	Level 2	Level 3*
1 Urban or built-up	11 Residential 12 Commercial and Services 13 Industrial 14 Transport, communications 15 Industrial and commercial complexes 16 Mixed urban or built-up land 17 Other urban or built-up land	111 Single Family 112 Multi-family 113 Group quarters 114 Residential hotels 115 Mobile home parks 116 Transient lodgings 117 Other
2 Agricultural land	21 Cropland and pasture 22 Orchards, groves, vineyards and nurseries 23 Confined feeding operations 24 Other agricultural land	
3 Rangeland	31 Herbaceous rangeland 32 Shrub and brush rangeland 33 Mixed rangeland	
4 Forest land	41 Deciduous forest 42 Coniferous forest 43 Mixed forest	
5 Water areas	51 Streams and canals 52 Lakes 53 Reservoirs 54 Bays and estuaries	
6 Wetland	61 Forested wetland 62 Non-forested wetland	
7 Barren land	71 Dry salt flats 72 Beaches 73 Sandy areas other than beaches 74 Bare, exposed rock 75 Stripmines, quarries and gravel pits 76 Transitional areas 77 Mixed barren land	
8 Tundra	81 Shrub and brush tundra 82 Herbaceous tundra 83 Bare ground tundra 84 Wet tundra 85 Mixed tundra	
9 Perennial snow or ice	91 Perennial snowfields 92 Glaciers	

* 11 Residential is broken down to level 3 as an example.

4.6.6 AN EXAMPLE OF THEMATIC MAPPING FROM LANDSAT DATA

A research project at the University of Aberdeen involved the development and testing of a methodology for the classification of land cover types in Scotland from Landsat digital data (Wright and Hubbard 1982). The project entailed the integrated use of ground sampling and aerial photographic interpretation to establish suitable 'training' data and to permit a semi-automated classification of land cover types for the seven Landsat scenes required to cover the mainland of Scotland (about 75 000 km²).

Initial problems of misclassification were encountered due to the great variation in reflectance values for the same land cover types within any one scene, due largely to atmospheric effects, ground relief and shadowing. This was overcome by dividing each scene into a number of sub-scenes, of fairly homogeneous terrain and atmospheric characteristics, and establishing 'training' data within each sub-scene prior to computer classification of the rest of the scene. Utilising scenes which had been geometrically corrected and resampled to 100 m pixel size (by the National Remote Sensing Centre, Farnborough), a data matrix of the seven scenes was created (removing the scene overlaps) in which the classified data for each sub-scene were stored.

The data for the classification of the whole area, or part of the area, can be used to produce separations of the blue/green/red components of a colour-coded version of the classification, on a high quality film-writer machine. These film separations, in turn, are used to produce separation plates for colour printing of maps of the land cover classification.

Since each pixel represents a 100 m square (or 1 hectare) on the ground, the area statistics for each land cover class can be easily abstracted. The regional administrative boundaries of Scotland were introduced into the data matrix so that statistics and maps of land cover types for each region may be abstracted separately. An example of this is given in Plate 8 for Tayside Region.

Accuracy checks on the classification indicated that the 85 per cent overall accuracy for all categories of land cover required by Anderson *et al.* (1976) had been achieved.

4.6.7 SOME LIMITATIONS OF LANDSAT

A brief acquaintance with Landsat data may produce a misleading impression of the capabilities of the system for mapping and monitoring the earth's surface features on a grand scale. It should be remembered that the basic MSS nominal ground resolution of 79 m was attained by Landsat 1 in 1972 and remained essentially the same for ten years.

The limitations of the Landsat system can be highlighted using the concept of *resolution*, as defined earlier in section 4.1.4. Improvements in the discriminatory capacity or in the classification accuracy attainable with a particular system are ultimately concerned with achieving a better resolution, be it spatial, spectral, radiometric or temporal resolution.

The *temporal* resolution of Landsat is 18 days, or 9 days in the case of Landsats 2 and 3 together, for some limited parts of the earth. For most applications a temporal resolution of 9, or even 18 days, is more than adequate, and in any case a shorter revisit time produces much more data, with consequent problems for handling and analysis. For many parts of the earth, however, 18 day repeat cycle is only theoretical, since the high incidence of cloud cover effectively prevents imaging in the Landsat wavebands (optical and near-infra-red). Each ground scene could be imaged 20 times per year, but there are significant differences in the imaging success rate for a good target area, such as the southwestern USA, and for a poor target area, such as the United Kingdom (Table 4.7). So, due to a combination of inadequate temporal resolution (shorter repeat cycle would have higher probability of imaging at more cloud-free times) and unsuitable spectral region of operation (clouds are opaque for visible and near-infra-red radiation), there are large areas of the globe which are unsuited to imaging by Landsat. Such areas should undoubtedly be better suited to a satellite-borne imaging radar system of the type planned for several imaging satellites of the 1990s.

For many users, *spatial* resolution of ground details is the most important resolution type. Line-scanner instruments cannot achieve a spatial resol-

Table 4.7 **United Kingdom and southwestern USA: Landsat scenes acquired with less than 30 per cent cloud cover during the period July 1972 to April 1976**

Region	Number of scenes acquired	Number less than 30 per cent cloud	Percentage of total scenes acquired
United Kingdom (51 frames)	390	79	20.3
Southwestern USA (51 frames)	4310	3044	70.7

(*Source:* NASA Goddard Space Flight Centre, 1979)

ution comparable with that of a camera/film system from a similar altitude. During the late 1970s, Landsat MSS produced nothing to compare with the 17 m nominal ground resolution photography from the Skylab Earth Terrain Camera (Table 4.5).

It is ironic that the region of best spatial resolution (optical wavebands) is where there tends to be greatest spectral ambiguity for natural materials, such as vegetation. A much greater range of natural spectral responses occurs outside the optical waveband interval. It is an enduring problem of remote sensing systems, therefore, that the *spatial* resolution is best where the *spectral* resolution is generally least useful.

Ground features cannot be classified unless they have first been discriminated, either spatially or spectrally, by the sensing system. On this basis the Landsat MSS system is of poor spatial resolution for the requirements of most field scientists, and has 4 spectral bands in a region (visible and near-infra-red) where spectral discrimination is not optimal for most features. There has been a considerable improvement with the Thematic Mapper on Landsat 4, with the nominal spatial resolution (IFOV) improved to 30 m (Fig. 4.21) and 7 spectral bands with three in significant regions of the infra-red spectrum not included on the MSS (Fig. 4.15).

Even 30 m spatial resolution is not adequate for many mapping requirements, and in some cases high-altitude colour infra-red aerial photography (with a nominal ground resolution of a few metres) may be digitised using a scanning microdensitometer, so that the digital values are amenable to processing, analysis and classification on the advanced image analysis systems originally developed for use with linescanner data.

Landsat linescanner systems in 1987 had produced data of no better than 30m nominal ground resolution, although it was technically possible for the Landsat 4 Thematic Mapper to be designed for a better (finer) spatial resolution. Satellites orbiting the earth can, however, image the terrain indiscriminately, which is not to the liking of many countries. As spatial resolution improves, then countries become highly sensitive to the security implications of such an imaging satellite.

4.6.8 SPOT

In February 1986 the French earth imaging satellite, SPOT (Systeme Probatoire pour l'Observation de la Terre), was launched into a sun-synchronous, near circum-polar orbit at 832 km altitude. The sensors are two identical high resolution visible (HRV) imagers which use a linear array of 6000 charged coupled detectors (CCD). The HRV instruments can be operated in either a 3-band multispectral XS mode (0.50–0.59, 0.61–0.68 and 0.79–0.89 μm) with a ground resolution of 20m, or in a single band panchromatic mode (PAN, 0.51–0.73 μm) with a 10 m ground resolution. The ground swath at nadir is 60 km.

For nadir viewing the repeat cycle of SPOT is 26 days. However, a novel feature is the possibility of altering the viewing angle by remote 'steering' of the plane mirror which directs the radiation from the ground onto the array of detectors. This permits 'off-nadir' viewing of the terrain at up to 27° from the vertical. It is thus possible for the same area to be viewed at intervals of 1,4 and sometimes 5 days.

SPOT also permits stereoscopic imagery to be acquired, by recording the same ground area on two successive days using the 'off-nadir' viewing capability. In preliminary tests carried out by the Institut Geographic National in France, a heighting precision of ± 5 m has been claimed for stereo-measurements.

Although 50,000 cloud-free scenes were recorded during the first year, only about 10 percent were available for use, thus highlighting the basic problem of ground processing of satellite digital data.

4.7 SATELLITE REMOTE SENSING: A SCENARIO FOR THE NEAR FUTURE

Improvements in spatial, spectral and temporal resolution would all be beneficial to further extending the applications of satellite remote sensing of the earth. However, improvements in all forms of resolution imply large increases in the data sets which have to be transmitted, processed, stored, accessed and analysed. If satellite remote sensing is to progress from an experimental to an operational phase, then there must be a greater proportion of investment devoted to the technology associated with storage, access and analysis of very large data sets. A possible future scenario has been recommended by the British Remote Sensing Society in its submission to the House of Lords Select Committee on Science and Technology. This recommends 'the establishment in the United Kingdom of a comprehensive national system for data storage and data preprocessing, connected to a series of regional nodes throughout the country via existing or planned computer networks. These nodes would offer access to remotely sensed data archives and to image processing facilities on a regional basis . . . .'

A significant problem for a centre-node network strategy would be the rapid transfer of data between the central facility and regional nodes . . . the development of METSATNET (operational 1984) shows that data transmission rates of 2 megabits per second can be achieved (i.e. one Landsat scene in 4 minutes). A likely candidate for wide dissemination of large sets of time dependent data is the technique called *controlled area broadcast* via satellite to small receive-only terminals' (Remote Sensing Society, 1983).

The analysis and classification of the remotely sensed data might then proceed using a regional digital image analysis facility or, at a more local level, within users' offices or laboratories, using a 'turnkey' image analysis system based on low-cost microprocessor technology.

FURTHER READING

Avery, T. E. and Berlin, G. L., 1985, *Interpretation of Aerial Photographs* (4th edn). Burgess, Minneapolis, USA.

Barrett, E. C. and Curtis, L. F., 1982, *Introduction to Environmental Remote Sensing* (2nd edn). Chapman and Hall.

Colwell, R. N., (editor-in-chief), 1983, *Manual of Remote Sensing* 2 vols (2nd edn). American Society of Photogrammetry and Remote Sensing, Falls Church, Virginia, USA.

Curran, P. J., 1985, *Principles of Remote Sensing* Longman.

Lillesand, T. M. and Kiefer, R. W., 1987, *Remote Sensing and Image Interpretation* (2nd edn) John Wiley.

Townshend, J. R. G., (ed.), 1981, *Terrain Analysis and Remote Sensing* George Allen and Unwin.

Ulaby, F. T., Moore, R. K. and Fung, A. K., 1981 *Microwave Remote Sensing: Active and Passive*: Vol. 1, *Fundamentals and Radiometry* Addison-Wesley, USA.

Chapter 5 CARTOGRAPHIC PRESENTATION

5.1 PRELIMINARY CONSIDERATIONS

It is often suggested, perhaps only by implication, that once the plot has been prepared the main task is over: all that remains is to 'draw the map in ink'. Stated this way, the importance of the cartographic stage is, to say the least, underestimated. In reality, this final process can have a profound effect on the quality of the work executed during measurement and data collection. Carelessness, poor design and layout, poor draughting and bad planning can considerably reduce the confidence the user should have in the final map.

In the following pages, the final field survey or photogrammetric plot is *not* considered to be the last stage in the whole process of mapmaking. After presenting an outline of the techniques of map production and reproduction, the methodology of map construction and design is described. While such a limited review is no substitute for training and experience, the reader must realise that he cannot satisfy the requirements of content, design and reproduction of his map without knowing some of the background to the whole process.

5.1.1 ASPECTS OF BASIC PLOTTING

As so much of field mapping and photogrammetric plotting is initially executed in pencil, some consideration is given here to this and other related techniques.

5.1.1.1 PENCILS

Lead pencils are normally used, different hardnesses being required for field and office plotting.

Pencils harder than 2H or 3H should be avoided since they can cause grooves in the drawing surface. Frequent sharpening is required to keep construction lines, etc., of uniform thickness, a round point and not a wedge being preferred. This is best carried out with a sharp penknife and then finished off by rubbing on a piece of paper – not on a sandpaper pad, since the dust produced is too easily transferred to the plot. Cleanliness should be maintained at all times. Pencils should not be used when they are less than half their length to maintain balance for efficient freehand drawing. Mechanical automatic-advance pencils with 0.3 mm leads may also be employed. Plastic erasers are ideal for most alterations, although a kneadable rubber may be useful at the final stage of cleaning the map.

5.1.1.2 OTHER INSTRUMENTS

There are many aids for the drawing of straight lines and curves: rulers and scales, set squares, tee squares, gridded film sheets, metal gridding plates, coordinatographs, protractors, French curves, railway curves, splines, compasses, etc. Readers are advised to experiment until they learn the potential of each instrument, and then to make the most appropriate use of it according to the work to be done.

5.1.1.3 SURFACES

The qualities important for basic field mapping or basic plotting are dimensional stability, erasing quality, strength and reaction to wetting. For field plotting, smooth good-quality cartridge paper, aluminium-based paper, and board or polyester draughting film are best. These surfaces are also suitable for office plotting. Note that much harder pencils are required for drawing on film than on paper, as film has a hard, mildly abrasive surface. It should be remembered also that paper and film come in standard cut sizes or can be bought off, or by, the roll, and rolls are of fixed widths (e.g. 76 cm and 102 cm).

5.1.1.4 PENCIL DRAWING

Most people think that they are able to draw with a pencil, but few can do it well since it takes training and expertise as well as a little talent. Lines should be firm, clean and even. Evenness may be achieved by twisting the pencil in the fingers as the line is being drawn, the angle between pencil and surface being maintained at about 80°. Practice should be obtained in the drawing of light lines and heavier lines, without grooving the paper; this skill is important when the field plot has to suffice as the final map. If lines should cross at a point, make them do so (Fig. 5.1), especially in plane tabling, where pressure of time, weather, etc., can weaken the resolve to draw with care.

Fig. 5.1 Drawing crossing lines.

The graphical construction of rectangular frameworks should be practised so that some of the aids mentioned above can be understood.

Freehand linework, frequently required in ground survey mapping, should appear just as if it had been drawn with a ruler – clean, firm, not sketchy. If long curved lines are to be drawn, for example, hold the pencil further back from the point than usual.

Care and cleanliness should be maintained on both field and office plots. Pencils and instruments should be kept in good condition, pencils not being sharpened over the drawing! Hands should be kept clean and out of contact with the drawing surface. Take care not to dirty or damage the surface by sliding instruments around: lift them clear whenever possible. Erasure should be done with care

and the resulting particles removed at once from the drawing. The cleaner and more complete a drawing is, the better it will be suited for inking up or later copying.

5.1.1.5 SYMBOLS AND ESSENTIAL INFORMATION FOR BASIC PLOTS

The design of symbols for the final map can be worked out in the office, but for plane-table work and for plotting tacheometric sheets, etc., it is useful to have some standard symbols or conventions. The maps in these instances are mainly line maps, with lines for the boundary framework, physical boundaries (e.g. fences), administrative boundaries, routes, contours, etc., and there may also be a number of point symbols. The main principle should be to avoid ambiguity; if in doubt, give an explanation in a key on the map sheet. The simplest solution might be to adopt the conventions used on the official large-scale maps, but do not hesitate to reconsider these symbols if necessary when it comes to the final drawing.

5.1.2 CHECKING THE FINAL PLOT

Before production of the final map, the plot must be tidied up. (If, as in photogrammetry, relief and culture have been produced on separate sheets of film there will of course be a number of plots.) Unwanted marks, lines, etc., are removed; names and symbols are clarified and errors are noted. Ideally, this should be done as soon as possible after surveying or photogrammetry has been completed, and preferably when there is still an opportunity to return to the field or photogrammetric model in order to remedy errors or omissions. Scale, orientation, location, date of survey and the surveyors' names should be noted on the map, along with the instruments and methods used in the mapping process. This information, too often omitted from maps, can benefit future users, as can some indication of variations in the quality or intensity of field mapping in particular. Since reference may have to be made to original field notes and sketches at this stage, full and legible field notes are most valuable.

5.1.3 PURPOSE OF THE END PRODUCT

The originator must have a clear idea of the purpose of the map, profile or cross-section, whichever may be deemed the most appropriate end product. Surprisingly, the nature of the end product does not affect the discussion, since all such images are composed of points, lines, areas and text, which must be designed and drawn in the usual way. Commonly, the task will be to present, in a topographic context, a particular distribution of features, boundaries, etc., permitting further measurements and analyses. The simplest requirement might be for a clean, checked basic plot, in pencil, as produced from the process described in the previous section. It is, however, advisable to make the manuscript more permanent, either by drawing in ink over the pencil lines, or, preferably, by tracing the map in ink on to a separate sheet of polyester draughting film. If the work is to be utilised by others, it should be designed and drawn with particular care. Crisp, neat work will enhance the content and give confidence to the user. This stage of redrawing the basic plots can be separated, as suggested earlier, e.g. one sheet carrying only the relief information while the other bears the planimetric detail. This is particularly desirable if a complex contour image must be combined with detailed planimetry. It can also facilitate later production procedures and is permissible if the sheets are carefully registered, i.e. matched accurately, detail for detail, to allow for perfect superimposition. However, such separation is quite unnecessary if the final product is to be a simple black and white map without the anticipation of colour printing, or more sophisticated methods of production. In this case, it may be necessary to produce a compilation-stage map, combining the material from the relief and planimetry plots, before drawing the map in ink on a final sheet of draughting film.

5.2 TECHNICAL ASPECTS OF MAP PRODUCTION: BACKGROUND NOTES

The field scientist may have to draw his own map (which is often preferable), but even if he is able to use the services of cartographers, he should be familiar with the main techniques and procedures of map construction and reproduction. He can then participate fully in the discussion and planning of his work or even decide whether or not he should carry out the final drawing himself. High-quality images must be produced, whatever the end product.

5.2.1 PRODUCING A CRISP, BLACK IMAGE

5.2.1.1 STANDARD TECHNIQUES

Drawing in ink can be enhanced by the employment of certain of the techniques and equipment described below.

Fig. 5.2 Tubular pen: Rotring 'Variant', complete and dismantled.

There are numerous instruments for drawing, but only the tubular pen with interchangeable point sizes (Fig. 5.2) is recommended here, both for its versatility in use and its ease of maintenance (e.g. Rotring, K + E, Standardgraph). Its main advantage is that the interchangeable points are designed to produce a free flow of ink and constant line widths. With a range of thicknesses from 0.1 mm upwards, it also provides for maximum scope in design. The ability to produce and maintain a thin line can be critical, especially if the survey is a precise one, where crude variations in line thickness could introduce scale errors.

Appropriate ink is normally recommended by the manufacturer of each pen, but it should be free-flowing and produce a good black image on the surface, paper or film, for which it is designed.

If the map is to be drawn directly on the plotting

sheet, the nature of this surface will partly determine the quality of the final image. Smooth, heavy quality papers and boards are easiest to use and permit the drawing of fine ink lines. Poorer papers may cause the ink to spread through the fibres, destroying any precision obtained at the pencil stage. Bad fieldwork practice could damage a paper plane-table map irreparably and lead to very imprecise results. Draughting-film field sheets, however, produce a better result. Polyester draughting film is normally translucent rather than transparent, is available in various thicknesses and is either matt on one side or, in the more opaque variety, on both sides. For work of the type discussed in this book double-matt material, 0.3 mm thick, would be adequate. While normal tubular technical pens can be used on this material, suitable recommended inks should be obtained. One advantage of draughting film is ease of erasure, normal inks being removed by a moistened stub and plastic eraser.

It should be confirmed that any such inks are chemically suited to the drawing instruments. Some inks are designed to etch into the plastic surface and should be avoided completely, as the etching process might cause the pen to disintegrate unless it is specially designed to accept such inks. Other inks, which are more advisable, may be more viscous, however, and should be cleaned from the pen immediately after use to avoid clogging.

Short periods of working should not affect the standard pen point, but abrasion is continuous, and for long-term work on draughting film special pens are required. (Refer to current catalogues.)

Although tracing paper is translucent, permits modest erasure and is inexpensive, it is not suitable for the production of large-scale maps because of its instability. Even in its thicker grades, it will distort differentially with the application of ink, and

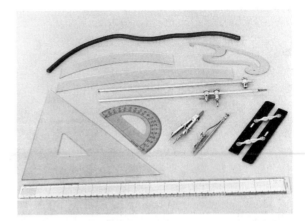

Fig. 5.3 Drawing aids: flexicurve, French curve, railway curves, beam compass, set square, protractor, spring bow compass, drop compass, parallel rule, and 45 cm rule.

under varying conditions of temperature and moisture. Deterioration is accelerated if the map passes through certain copying processes (e.g. diazo); and if colour-separated originals are to be drawn to permit later colour printing, tracing paper does not provide for the exact registration of detail. If, however, these rigorous requirements need not be met, tracing paper may be employed.

Instruments designed to improve the drawing of straight lines, curved lines, rectangular frameworks, etc., are illustrated in Fig. 5.3. Other linework must be drawn freehand, and skill will only come after training and experience. Although point symbols can be drawn freehand, or with the aid of straight edges, curves or compasses, symbol stencils may also be employed. When a stencil or ruler is being used with ink, the working edge of the guide should be raised just clear of the surface to avoid surplus ink spreading under it and ruining the work (Fig. 5.4). Small patches of draughting tape fixed to the undersides, set slightly in from the edge,

Fig. 5.4 Preventing ink from spreading beneath drawing guides.

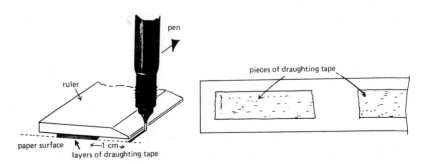

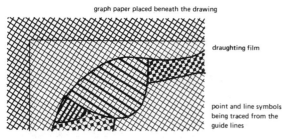

graph paper placed beneath the drawing

draughting film

point and line symbols being traced from the guide lines

Fig. 5.5 Use of graph paper as guide to regular patterns.

should suffice. Area symbols are most simply produced by repeating point or line symbols in either regular or irregular patterns, and in combination. Regular patterns of lines or evenly spaced points can be located by measurement on the surface, or by using special ruling instruments. Where the plot is being redrawn on draughting film, graph paper (placed behind the film) can be used as a guide (Fig. 5.5). Sheets of adhesive symbols designed to accelerate the production of highly professional artwork are available in point, area and line form, in a wide variety of designs from manufacturers such as Letraset and Meca-

norma (Fig. 5.6). Their use, however, can add significantly to the cost of the final drawing (Table 5.1). Adhesive symbols do not stick firmly to tracing paper, and area symbols especially (e.g. Letratone) may show wrinkles as the tracing paper changes slightly in size or if the map is rolled. Surfaces such as Bristol board and draughting film, however, do not suffer from all these problems.

Freehand lettering, if carefully planned and executed, may be appropriate, but otherwise stencils or lettering guides should be used (Fig. 5.7). It was in this area that adhesive products first excelled. Although, strictly speaking, trade names, the words 'Letraset' and 'Artype' are now almost synonymous with the technique they represent, and the variety of styles and sizes produced in such systems can meet most requirements (Fig. 5.8). One advantage of an image produced entirely in pen and ink, however, is its completeness, uniformity and durability. Adhesive symbols and lettering, on the other hand, produce final images that are rather more vulnerable, but special protective spray coatings (e.g. Letraset 101) reduce the risk of damage, for example, during passage

Fig. 5.6 Adhesive area symbols.

146

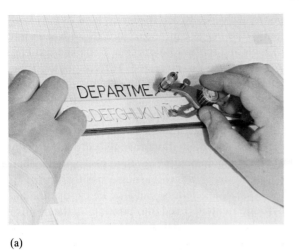

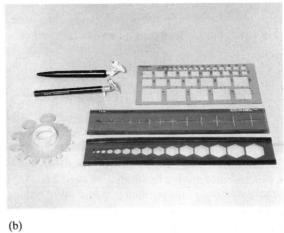

(a)   (b)

Fig. 5.7   Aids for lettering and symbols: lettering guide (a), plastic stencils and stencil pens.

Letra   **Letra**   **Letraset** *instant lettering*   14pt HELVETICA MEDIUM

A A A A A A A A	A A A A A A
A A A A A A A A	A A A A A B
C C C C C C C C C	C C C D D D
D D D E E E E E	F F F F F F F
E E E E E E E E	F F F F F F F
F F F F F F F F F	G G G G G G
H H H H H H H	H H H H H H H
I I I I I I I I I I I I	I I K K K K K
J J K K K K K K I	L L L L L M M
L L L L L L L M M	N N N N N N
N N N N N N N I	N N N N N N
N N N N N N N	O O O O O O
O O O O O O O O O	Q Q R R R R R
P P P P P Q Q Q Q	R S S S S S S S
R R R R R R R R	T T T T T T
S S S S S S S S S	T T T U U U U
T T T T T T T T T	U U U V V V
T T T T T T U U U	W W X X X X
U U U U U U V V	
W W W W X X X X	

AAAAAAAAAAAAAAAAAAAAAAAA   ääääàaaaaaaaaaaaaaaaa
AAAABBBBBBBBBBCCCCCCCCC;   abbbbbbbbbbcccccccccc
CCCDDDDDDDDDDDDDDDEEE;   ddddddddddèééeeeee
EEEEEEEEEEEEEEEEEEEEEEE   eeeeeeeeeeeeeeeeeee
EEEEEEEEEEEEEEEFFFFFFFFFFFF   eeffffffffffggggggggggg
GGGGGGGGGGGGGGHHHHHHHHH;   hhhhhhhhhhhhhhiiiiiiiii
HHHHHHHHHIIIIIIIIIIIIIIIIIIIIIIII;   iiiiiijjjjkkkkkkklllllllllll
IIIJJJJJKKKKKKKKLLLLLLLLLLLLL   mmmmmmmmmmmmnnn
LLLLLLLLMMMMMMMMMMMMNN   nnnnnnnnnnnnnnnnnöö
NNNNNNNNNNNNNNNNNNNN;   ooooooooooooopppp
NNNNOOOOOOOOOOOOOOOO   ppqqqqqqqqrrrrrrrrrrrr
OOOOOOPPPPPPPPPPPPQQQQQ   ssssssssssssssssssss
QQRRRRRRRRRRRRRRRRRRRR   tttttttttttttttttttttttttü
RSSSSSSSSSSSSSSSSSSSSSS;   uuuuuuuuuuuuvvvvvv
SSSSSTTTTTTTTTTTTTTTTTTT   wwwwwwxxxxyyyyyyyy2
TTTTTTTUUUUUUUUUUUUUUU   1111112222233333444
UUUUUUVVVVVVVVWWWWWWWW   55566666777777788888
WWXXXXYYYYYYYYZZZZZZZZ%;   99000000&&??!!ßß££$

AAAAAAAAAAAAAAAAAAAAAAA   ääääàaaaaaaaaaaaaaaaa
AAAABBBBBBBBBBCCCCCCCCC;   abbbbbbbbbbcccccccccc
CCCDDDDDDDDDDDDDDDEEE;   ddddddddddèééeeeee
EEEEEEEEEEEEEEEEEEEEEEEE   eeeeeeeeeeeeeeeeeee

Fig. 5.8   Adhesive lettering.

through copying processes, when the map image may be subjected to heat, light, moisture, distortion and pressure.

5.2.1.2 SCRIBING AND RELATED TECHNIQUES

The more important and extensive the survey and the more demanding the potential use, the less inclined will the field scientist be to do his own cartographic work. In such cases, the cartographers employed may prefer to adopt a method known as negative scribing, rather than pen and ink. Scribing produces a negative image by cutting lines, etc., in the opaque coated surface of a transparent plastic sheet. When this line negative is printed in contact with photographic film or paper, a perfect black image results.

Scribing, now used universally by professional mapmakers, is becoming more popular in smaller drawing offices. One significant advantage is that high-quality linework can be produced by fairly inexperienced personnel. If the basic field manuscript image is transferred to the scribing surface photomechanically, or is merely used as a tracing guide, the image can be cut directly with a variety of tools and scribing points (Fig. 5.9). The resulting negative can be corrected before photographic contact prints are made. Symbols and lettering, not added at the negative stage, can be fixed to this positive or produced separately by means of photo-setting techniques, in which words and symbols are prepared and printed photographically by specially designed equipment. The thin photographic film (stripping film), which carries a positive image of the symbols and letters, can later be cut up, and the appropriate items fixed into position on the positive map image by special cement or wax (Fig. 5.10). Small microprocessor controlled lettering machines are now quite common and provide a lower cost, slightly lower quality alternative. (e.g. Kroy, 3M).

Finally, if very fine area patterns are required, 'peel coat' can be employed; from this, areas can be cut and peeled away, leaving open 'windows'. If these 'masks' are placed in direct contact, along with the linework, etc., with the final photographic film, they produce areas of solid black to correspond with each window. If, however, a negative

Fig. 5.9 Scribing.

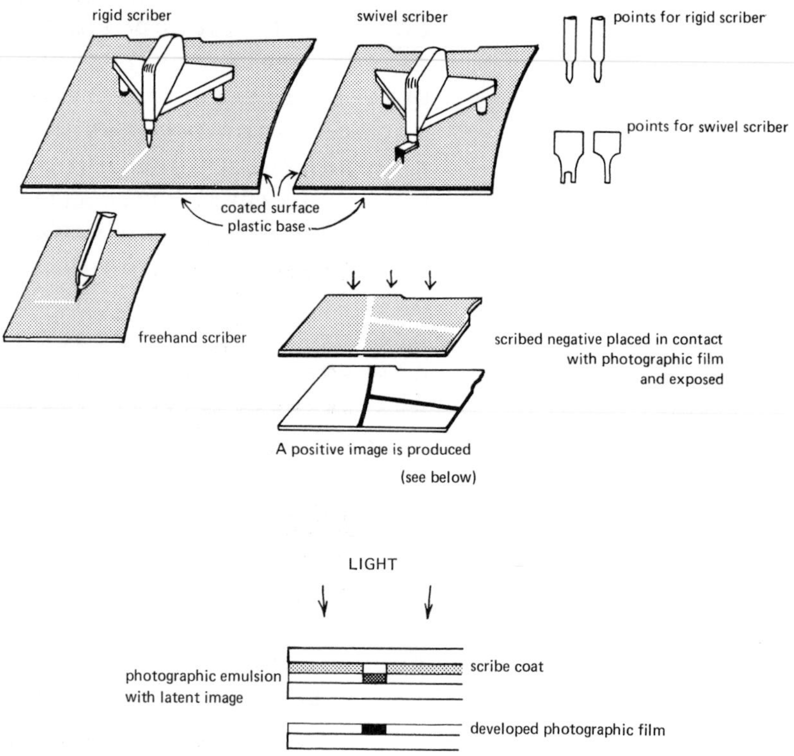

148

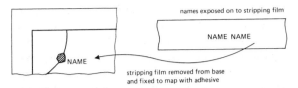

names exposed on to stripping film

NAME NAME

NAME

stripping film removed from base
and fixed to map with adhesive

Fig. 5.10 The use of filmset lettering for maps.

halftone tint screen is interposed at the final exposure stage, a 'grey' tint effect will be produced instead. Other patterns may also be used (Fig. 5.11). Standard tints are available from most reprographic suppliers.

Naturally, the decision to use scribing will depend on the equipment and expertise available. The larger and more important the survey, the proportionally cheaper the materials become for this method. Furthermore, it has been found that the dense black image produced in this way is less likely to cause the reproduction problems photo-

graphers and printers encounter with poor quality 'grey–black' ink lines which can break up too easily.

5.2.1.3 COMPARATIVE COSTS AND QUALITY

The comparative costs of equipment and materials are given for general reference in Table 5.1. It should be emphasised that if one abandons the simple ink drawing, either to include adhesive symbols and lettering, or ultimately to employ the scribing technique, the cost increases, as does the time required and the dependence on skilled personnel and specialised photographic and reprographic facilities, which may lie beyond the control of the draughtsman. If such facilities are available, the more sophisticated methods may be simple to adopt – if they are seen to be necessary. If any part of the map production must be completed by

Fig. 5.11 The stripmasking technique for area symbols.

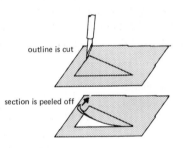

outline is cut

section is peeled off

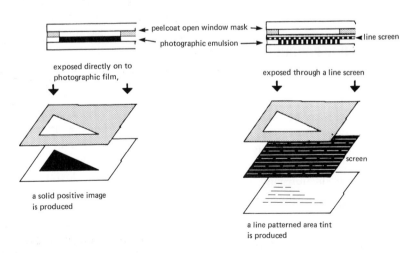

peelcoat open window mask
photographic emulsion
line screen

exposed directly on to photographic film,

exposed through a line screen

a solid positive image is produced

screen

a line patterned area tint is produced

NB:
These exposures would be made with the component sheets in direct contact.

Table 5.1 Comparative costs of consumable materials for a single-colour map (Terms are not fully explained if the process indicated lies beyond the scope of this book.)

Method 1 *Freehand linework/stencils, etc., on field plot*
    1 bottle ink
    1 sheet cartridge paper (50 × 75 cm)

                                        total: x

Method 2 *As above, on 0.3 mm gauge double-matt draughting film*
    1 bottle plastic ink
    1 sheet draughting film (50 × 75 cm)

                                        total: 4x

Method 3 *Planned on draughting film and drawn separately on film using adhesive symbols*
    1 bottle plastic ink
    2 sheets draughting film (50 × 75 cm)
    3 sheets Letraset
    1 sheet Letraset symbols
    2 sheets Letratone
    1 can protective spray

                                        total: 22x

Method 4 *Scribing and attendant photographic processes*
    1 sheet draughting film for planning (50 × 75 cm)
    1 sheet scribing film
    2 sheets peel-coat mask
    2 sheets disposable negative area tints
    photoset 35 mm film
    3 sheets photo-contact film
    correcting fluid and developer

                                        total: 25x

---

another agency, however, the effects of possible delay or breakdown in communication should be foreseen.

5.2.2 MAP REPRODUCTION

It may be necessary to copy a map once or several hundred times. In designing the map, therefore, such a need must be anticipated, since the design and quality of the image affect the nature and number of copies that may be made. Only a few of the wide range of techniques are described here.

Map reproduction may be divided broadly into 'non-printing' (copying) and 'printing' methods.

5.2.2.1 NON-PRINTING METHODS

Although the field plot may seldom require copying, there are sometimes valid reasons for doing so. Foresight in the production of a good final image facilitates the provision of copies. The copying process is more suited to small numbers of reproductions. If the production of each copy employs a procedure that must be repeated almost entirely each time, there will be little or no reduction in cost per copy as numbers increase. Facilities for both same-scale and reduced-scale copies are available, but the former are being considered here.

Photography

Most commercial photocopying devices are suited to smaller formats only (e.g. A3, A4). If copies are required of larger originals, access to special contact copying facilities is necessary. A good black image on translucent/transparent draughting film is best for this process, although reflex copies can be made from originals on opaque bases. Various photographic materials can be employed, and copies made on film or paper, some grades of the latter rendering an image not unlike conventional printing. All copies are permanent. These processes, however, can be quite expensive, since a high-quality negative may have to be made as an intermediate stage, and the whole process requires skilled technicians.

Diazo

Here the copying surface (paper, film, etc.) is coated with a diazonium compound. The original, which must at least be translucent, is placed in contact with it, either around a roller or in a flat vacuum contact frame, and exposed to u–v rich light, which passes through the original on to the sensitised paper, thus destroying the diazonium. Except where protected by the positive image, a latent image remains and this is normally developed dry, a process in which the developing complex is in the paper and is activated by exposure to ammonia gas.

The blacker and crisper the original image, and the clearer the base, then the better is the copy. Large copies are easy to make and relatively cheap in materials and labour. None of these copies, however, is permanent and will fade with prolonged exposure to light. It is advisable to put a clear warning of this eventuality on the edge of the map sheet.

Other processes, such as electrostatic copying, are available, but machines tend to have limited format and seldom produce copies larger than A3.

5.2.2.2 PRINTING PROCESSES

In these processes the original image is transferred to a printing surface, e.g. a printing plate, which can be used repeatedly to transfer the image on to paper sheets. Initial costs for equipment and plate-making are high, but the cost per copy reduces rapidly as the number of printed impressions is increased. Three major printing processes are normally distinguished.

1. Intaglio: copper engraving in the past, photo-gravure today.
2. Relief: the woodcut of the early days, letterpress of today.
3. Planographic: lithography.
4. Stencil: screen printing.

The first two are not important in modern mapmaking. Offset lithography is now the dominant method for the reproduction of maps. Normally, the printing plate is made from a wrong-reading (mirror reversed) final negative or positive. This plate (right-reading) is then fixed to a rotating cylinder in the printing machine, which successively coats the plate image surface with printing ink and then brings it into contact with the printing paper via an intermediate rubber cylinder ('rubber blanket'), i.e. the printing plate image is offset on to this rubber blanket and then on to the paper. The quality of the copies is high and equipment ranges from single-colour A4 small-offset machinery to massive four- and six-colour printing machines which produce images measuring 100 cm × 150 cm and larger. Small-offset lithography, employing cheap plates, may be economical for small format work down to twenty-five copies, although quality may suffer slightly. Screen printing can be used to produce smaller numbers of copies of coloured maps more cheaply than by lithography. It can also be used to add area colours to a map previously printed in black outline. For present requirements the important considerations are cost and deciding when a particular process should be used; these are best discussed with a professional printer in relation to each job.

5.3 MAP DESIGN AND CONSTRUCTION

It is obvious that the design and construction of a map will precede its reproduction, but in this book, reprographics has been considered first because design and construction will be carried out in the knowledge that certain drawing equipment and reproduction processes are available, such considerations almost certainly being built into the original specification. It is against this technical setting that the design and construction of the map can now proceed. Design is the intellectual decision-making stage, while construction is the translation of the design into concrete image form. Although the whole process of map production is a continuous one, it is convenient to break it down into stages.

5.3.1 CONTEXT AND REQUIREMENTS

The context of the map (for example whether it is to be a sheet map or is to appear on a page in a report) will influence the general design requirements and should be reviewed and verified. The design requirements should comprise scale, possible uses of the map, number of copies required and specifications of size, shape and general style. Such a specification might take the following form.

Format: Determined by the scale selected and the area depicted.

Style: Black ink only. Symbols drawn with precision, clearly contrasted but without particular emphasis on any one category. Lettering to be uniform, legible but not too bold.

Reproduction: Diazo copies only – hence the final drawing should be on a transluscent surface and be suitably protected.

Construction: Drawn with ink on draughting film with the addition of adhesive lettering and area symbols where required.

If maps are being prepared for publication in journals, etc., the author should seek the advice of the editor, obtain a copy of the relevant guide for authors and then discuss the matter with a cartographer. This is particularly important if, as is common, the map may subsequently have to be reduced in size for publication.

5.3.2 COMPILATION

The basic compilation is normally in existance at the completion of the final plot of the original survey. It only remains to decide on the possible omission of some features and the inclusion of certain additional items, such as grid lines, grid edge ticks, marginal information and, of course, names. Marginal information will comprise title, sheet number, scale style, north point, explanation of symbols, date of survey, grid data, index to adjoining sheets (if any exist), location diagram, reliability diagram (indicating any differences in the quality of the survey work from place to place over the map), surveyor/author, source base maps used and acknowledgements for aid (financial or otherwise). The exact nature and wording of these data should be specified, keeping them as brief as possible. Credits or acknowledgements should be simple, referring to organisations rather than to the lists of individuals which are more appropriate in written work.

Although the compilation may appear to be complete, a more analytical study of the compilation process will reveal the following stages for consideration in the event of a need to modify the map for alternative presentations.

5.3.2.1 GENERALISATION

In the normal process of producing the final drawing of a field plot at the original survey scale, generalisation would not be considered. In the event of modifying the original map – for a smaller scale, for example – it could become necessary. In this case, two important controls come into force: map purpose, which determines the degree of generalisation desired; and the scale to which it will be reduced, which may necessitate generalisation in order to preserve legibility (Fig. 5.12).

Generalisation should begin with selection of the information to be included or, alternatively, the removal of the least important elements (Fig. 5.13).

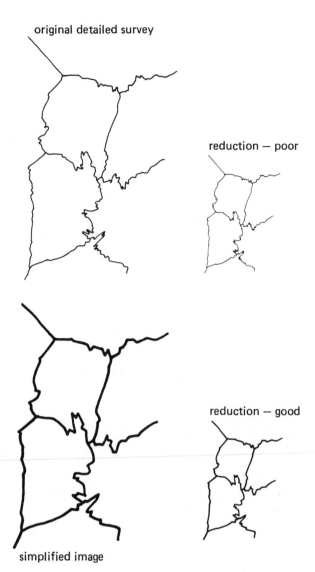

original detailed survey

reduction – poor

reduction – good

simplified image

Fig. 5.12   Some examples of generalisation.

Classification could be introduced, e.g. one type of field boundary instead of three. Simplification of redundant detail may be required for more intricate areas such as complex woodland, although it is often surprising how much information can be retained without modification, even when the scale is reduced to half. Essential information, such as main roads, boundaries and contours, should not be modified if it is possible to retain it.

If simplification is essential, it may be achieved by exaggeration of the dimensions of a feature if it is important, by combination of smaller separate

Fig. 5.13 Generalising of field plot.

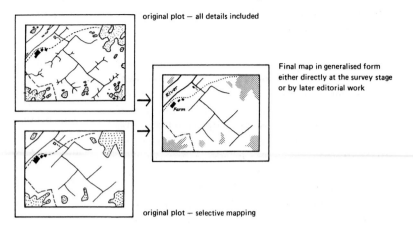

original plot — all details included

Final map in generalised form
either directly at the survey stage
or by later editorial work

original plot — selective mapping

elements or by omission of further detailed parts of the image. These strategies may also, in turn, give rise to displacement of some other features in certain areas where they are close together in reality. In all cases, the generalisation should preserve as much of the original information as is required by the new map and should correctly bring out the character of the remainder.

5.3.2.2 CONTENT EDITING

A reappraisal of the balance and the validity of the content is useful at this stage. The purpose should be considered, particular attention being paid to major primary users and using situations.

5.3.3 GRAPHIC DESIGN

'Design', a much misused term, covers in reality all the decision making which surrounds the production of anything, including a map. In the latter it covers content selection and the methods of production, and thus should not be restricted merely to graphic aspects such as planning the layout and choosing symbols. It is, however, the latter, more restricted interpretation that is considered below. Throughout the ensuing stages the reader should recall the technical aspects of map production, since any design decision must be made in the knowledge that certain tools and materials are available, and that restrictions could affect the final specification.

5.3.3.1 SOME THEORETICAL BASES FOR THE PRACTICE OF MAP DESIGN

In recent years the map design process has come under some scrutiny both from practising cartographers and from research investigators. Studies in visual perception, cognition and the philosophy of communication have provided some useful foundations, although the major problems are still far from solved. Readers may follow these studies of their own accord through selected references (see Bibliography) but, acknowledging that some useful guidelines may have emerged already, there follows a loosely structured background on the subject against which can be set the remaining account of the practical cartographic process.

Symbols are grouped primarily as points, lines and areas, and most symbol analyses accept this subdivision. It is suggested, also, that six so-called visual variables can be applied to these basic symbol groups either singly or in combination: size, lightness value, grain (textural gradient), colour, orientation and form (shape).

There are two important ways in which graphic symbols can be judged.

1. *The first graphic impression of the symbols*
   Here the primary impact needs no explanation. The message of the symbol scheme is perceived instantly and is understood at that fundamental level. The relationships are as follows.

   a. *Nominal or differentiation scaling*
      Similar symbols represent features that are similar, while different symbols mean differ-

ences in reality. The difference contrasts between these two groups may be as slight or as strong as the situation demands. All the visual variables can be employed to bring out this property of differentiation on their own, although the effect is often weakest with 'form' and strongest with 'size'. However, if the aspects of size or value should not be required then other variables should be selected.

b. *Ordinal scaling*

An ordering in the graphic scheme implies an ordering of features in reality. The impression of order is most vividly acquired by using the variables of 'size', lightness 'value' or 'grain' (textural gradient), although weaker effects can be achieved through certain applications of 'orientation' or 'form'.

c. *Relative proportions on an interval scale*

Symbols are adjusted to provide an instant impression not only of order but of relative proportions between the things portrayed. This can only be achieved directly through the variable 'size'.

This is not a comprehensive set of infallible rules but is merely a guide. These effects will often not work with very small point symbols or thin line symbols where the space is insufficient for the variable steps to become evident, e.g. square, circle, triangle 1 mm in size (Figs. 5.14 and 5.15).

2. *The meaning of the symbols*

Here, the symbol requires some explanation, although familiarity with context, conventions, etc. may help to enhance the direct effectiveness of the symbol design. A hospital symbol from

Fig. 5.14 The visual variables of graphics.

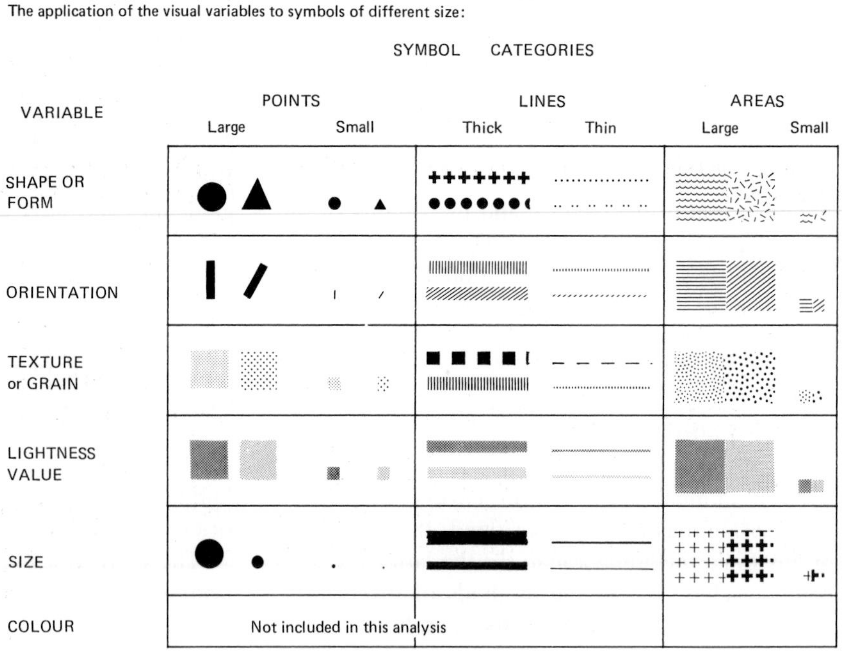

The application of the visual variables to symbols of different size:

SYMBOL CATEGORIES

This graphic table allows certain judgement to be made of the most appropriate design for simple symbols.

eg: shape variation alone in very small symbols is visually ineffective.

or, violent extremes of texture variation can be visually disturbing and render small areas difficult to portray,

some of these variables can be used individually (eg: size, colour) but combinations are more common

Individual judgement should be used to select the most appropriate and successful applications of these variables in a particular situation.

selected examples:

simple contrast effect:

weak, shape ○ △ □
strong, size ● ● ●

Good contrast effect (avoiding the magnitude implications of size and value):

colour [r] [g] [y]
orientation ▲ ◀ ►

The impression of order:

weak, shape □ △ ○
strong, size . ● ●
value
grain ‖‖‖‖ ‖‖‖ ‖‖‖

The impression of relative proportions:

weak, shape + ○ △
strong, size ● ● .

These are often combined in complex ways, the effects of variables not always being anticipated.
The correct effects of individual variables must be planned for. These effects may also be reinforced by selecting appropriate combinations, eg:

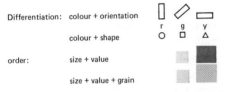

Differentiation: colour + orientation
r g y
colour + shape ○ □ △

order: size + value

size + value + grain

Fig. 5.15 Combining the visual variables for particular graphic effects.

the legend of an unfamiliar foreign map may cause confusion in the reader even if the cartographer feels that it is graphically closer to what is being represented (Fig. 5.16). This is an example of how the basic realisation that the cognitive map knowledge of potential users can influence the design process. Familiar symbols from the national maps of the country concerned may seem almost transparent. Their meaning is understood without any apparent visual analysis.

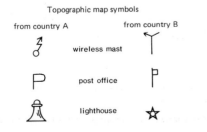

Topographic map symbols

from country A   from country B

wireless mast

post office

lighthouse

The normal users of map A will be less familiar with the symbols in map B.

Fig. 5.16 Selected symbols from topographic maps.

Building up symbol ideas

church   church
with spire

church
with tower

Fig. 5.17 The construction of a symbol.

Obviously the elements in (1) and (2) may be combined, but the map designer must first of all select specific symbol forms to portray the features on the map. This is done by considering semantics or, more simply, establishing the meanings required from the symbol, e.g. a building: composition? (stone); appearance? (shape); state? (ruined) (see Keates 1972 and 1982). Symbol ideas can be built up (Fig. 5.17). There are also situations where symbols should not be considered in isolation but their effects on one another taken into account. This applies to relief representation, for example, where the information contained in separate symbols such as contours, rock drawing, scree, etc. must also merge comfortably to create the grouped symbolic concepts of hills, valleys, etc. (Fig. 5.18). More basically, it may also be useful to consider certain overall relationships such as adjusting the symbol graphically to reflect the importance and size of a feature in the landscape, e.g.: continuous lines for strong barriers or physical boundaries such as walls, coastline, river bank and broken lines; or dotted lines for other features of lesser importance or which are less obvious in the landscape, such as footpaths and administrative boundaries (Fig. 5.19). Similarly, differentiation between strong, continuous symbols and broken or dotted symbols can also be employed to help reflect the precision of surveying, e.g.: continuous lines for surveyed contours; and broken lines for interpolated contours (Fig. 5.20). The major aim of such a

combining contours, cliff and scree:

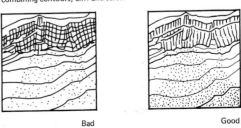

Bad   Good

Fig. 5.18 Combination of relief symbols.

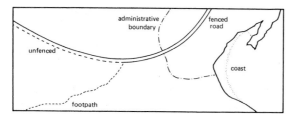

Fig. 5.19 **Symbols: strong and weak.**

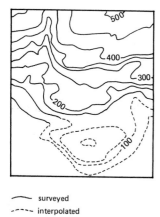

— surveyed
---- interpolated

Fig. 5.20 **Symbols: accuracy and approximation.**

strategy is that when someone picks up the map its appearance should create a subconscious awareness of the features that are going to be large, important and precisely surveyed and small, insignificant and perhaps only partially surveyed (Fig. 5.21). Once all these design criteria have been considered, the symbols for the whole map can be designed to satisfy the details and overall impressions desired.

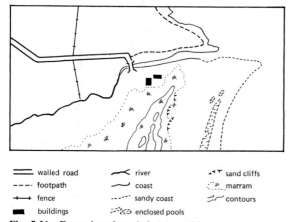

═══ walled road   ⌐ river   ⌃⌃ sand cliffs
---- footpath   ⌐ coast   ⁚⁚ marram
+—+ fence   ---- sandy coast   contours
■■ buildings   enclosed pools

Fig. 5.21 **Examples of symbols 'strength'.**

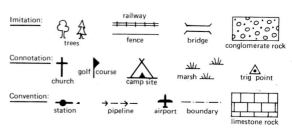

Fig. 5.22 **The forms of symbols.**

Perhaps most potential users will find it easier to understand familiar conventional symbols. If not, then the designer can return to the problem and can select either pictoral symbols – related to their referents by simple imitation or through connotation – or arbitrary symbols of an abstract geometrical nature (Fig. 5.22).

One of the common problems of many maps is that they can, in places, become too crowded with information and it may not be possible to reduce this clutter by generalisation alone. A slight rearrangement of names may help, but another strategy is available: separating related groups of symbols into imaginary layers or visual planes and attempting, through the use of visual depth cues, to portray them graphically as lying at increasing distances from the map reader! Although this technique can be applied in sophisticated ways there are many simple cases where two distinct graphical or visual planes of symbols may improve legibility, e.g. any thematic topic (such as geomorphology) superimposed on a topographic base map. If the map is to be printed in black only, then the symbols and names associated with the two layers are listed separately and the specifications adjusted to produce the desired effect. Superimposition is one common obvious effect, but other visual variables can be employed, e.g.:

Foreground: thick lines, larger point symbols, solid black symbols and sharply outlined area symbols.
Background: thinner lines, smaller point symbols, grey tones and areas with soft boundaries.

Names can also be adjusted to each level in the same way (Fig. 5.23). An easy way to achieve this overall effect is to arrange to have the base map printed in conjunction with a tint screen so that all the symbols are rendered in a grey tone, thus

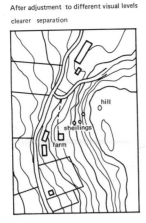

Before adjustment to visual levels
confused

After adjustment to different visual levels
clearer separation

Fig. 5.23   Visual levels in a map.

After screening of base map

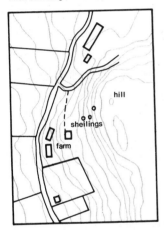

Fig. 5.24   Improving legibility by use of a tint screen.

removing confusion with the overlapping thematic symbols (Fig. 5.24).

This section is offered as a thinking framework and to indicate that all good design need not be an intuitive and inspired process. It is possible to set decisions against a background of gradually expanding useful theory.

5.3.3.2 A LOGICAL APPROACH TO THE DESIGN PROBLEM

Faced with an edited compilation sheet it should be possible to imagine how the finished map will look, mentally fitting the frame and considering where the title might best be placed. However, in acknowledgement of the foregoing section, attention should be concentrated on the features of the map itself, and how they can be converted from pencilled symbols, some of which may still require modification, into a professional image.

The first step is to list the features to be included in the map. Apart from presenting what must be symbolised, the list will also act as a check on omissions and as a model for the key when it is being prepared. The features that appear in Table 5.2 are not comprehensive, but are typical of the information contained in large-scale, specialised topographic maps. Many of these categories may be subdivided, thus extending the key where this is required.

Table 5.2   Typical features to be included in a large-scale topographic map

Buildings	inhabited
	uninhabited
	ruined
Field boundaries	fence
	wall
	hedge
Communications	railway
	metalled road
	track
	footpath
Water features	sea/lake
	river
	ditch
Vegetation/cover	bare ground
	scrub
	grass
	woodland
Contour interval	normal
	index
Spot heights	unmarked
	marked control points
Grid intersections	

5.3.3.3 SELECTION OF SYMBOLS

Bearing in mind the theoretical framework provided in sect. 5.3.3.2 above, a constructive 'doodling' stage should follow, as the designer thinks graphically about the mapped features (perhaps referring to field notes or even to photographs and other large-scale maps) and tests

feature	symbol design	modified for map
trees		
fence		
footpath		
building		ruined

map

Fig. 5.25   Adjusting the symbols to suit the map.

LEGIBILITY

poor – too small

good

poor – too intricate

good

CONTRAST

footpath boundary } poor contrast

foot path boundary } good contrast

Fig. 5.26   Legibility and contrast.

	0.05 mm
	0.25 mm space
	0.25 mm gap
	0.3 mm length of side
	0.3 mm circle diameter
	0.15 mm point diameter
	1.00 mm length of side
	0.1 mm dot spacing
	5.0 mm^2 minimum size of a patterned area

(drawings enlarged and not to scale)

Fig. 5.27   Selected accepted minimum dimensions for good legibility.

various designs and combinations on separate sheets of paper. These trials should be kept together for reference when the final decision is taken to begin drawing. As stated, this doodling should be constructive and follow some logical sequence. The exact form of each symbol should be selected, taking into account all aspects by which the feature is most commonly identified on the ground, or which the field scientist wishes to emphasise.

Once the basic symbol has been selected it can be modified according to its size, importance and condition (Fig. 5.25). At larger scales, it is less necessary to devise special symbols since most ground features on large-scale maps can be represented in detail, to scale, and named.

5.3.3.4 LEGIBILITY AND CONTRAST

The most important considerations in symbol design are legibility – symbols should be neither too small nor too intricate – and contrast – there should be sufficient difference between symbols to prevent confusion (Fig. 5.26). It is useful to be aware of the generally recognised minimum dimensions of various symbols for acceptable legibility (Fig. 5.27). It may often be adequate and more convenient to employ accepted conventional symbols as used in topographic map series of comparative scales, but these may not be ideally suited to the particular subject being mapped by the field scientist. It is therefore helpful to seek a logical approach to

symbol design, such as described above. It is better to understand some principles and procedures than merely to copy at random the symbolism used in other maps.

5.3.3.5 THE PROBLEM OF RELIEF

Some surveys produce detailed contours, at vertical intervals that may be as small as 1 m, plus numerous spot heights. To be of any value, contours must be easy to read, and this necessitates adjusting them in various ways such as varying their thickness and style or reducing their total number (Fig. 5.28).

contour interval   1m

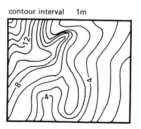

Fig. 5.28

158

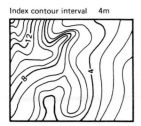

Index contour interval   4m

Fig. 5.29

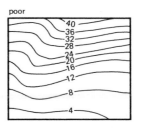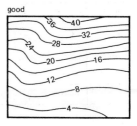

Fig. 5.31   Contour numbering.

The thickening of certain contours, selected at a wider vertical interval than on the plot, can accelerate the visual counting process to be carried out by the reader. Given a 1 m main interval, this new interval could be set at any size linked to the total range, e.g. every 3, 4, 5 or even 10 m. The selected contours are known as index contours (Fig. 5.29).

A common problem emerges in areas of extreme changes in gradient, i.e. wide, flat land with regions of very steep slopes. If a single narrow interval is adopted to suit the flat land, it may be too narrow for steep slopes and can cause bunching and coalescence of the lines. If the opposite is done, and a wider interval is selected to suit the steeper relief, the flat land may have to be depicted by one contour line only! A solution is to employ intermediate contours selectively, and this can be done in a variety of ways, e.g. a major interval of, say, 4 m could be selected to suit the steeper slopes, the 2 m or even the 1 m contours being added where necessary (Fig. 5.30). It is important to reduce the emphasis on these lines, so that, when the map is viewed normally, only the pattern of main contours stands out and a general impression of the whole landscape is gained. Closer inspection should bring out the details of the intermediate contours. The values of contours must be easy to identify, and hence contours should be clearly labelled across the map. The values should always be made to read

uphill; they should be inserted into breaks in the line and be scattered at a fairly even density across the map. Extended 'ladders' of contour numbers should be avoided (Fig. 5.31), since they introduce distracting and irrelevant lines into the contour image.

Spot heights, which should not be too prominent, can be located on isolated summits or in any areas of local ambiguity. If isolated closed contours occur frequently, as in sand dunes where no drainage features exist, a conventional symbol can be adopted to indicate whether they contain summits or hollows (Fig. 5.32).

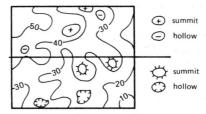

Fig. 5.32   Isolated, closed contours.

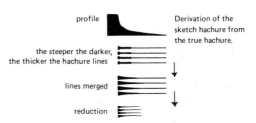

Fig. 5.33   Hachures.

intermediate contours

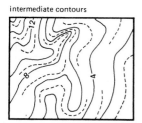

Fig. 5.30

In detailed relief mapping, contours are seldom suited to the depiction of all the landforms, and it is wise to employ a variety of other symbols. Many of these are basically sketch hachures of various forms (Fig. 5.33) which can, if desired, be modified for maximum three-dimensional effect by adjusting

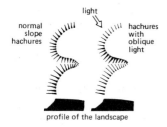

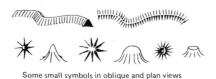

Fig. 5.34 Hachures with oblique light.

Fig. 5.36 Combined symbols.

Some small symbols in oblique and plan views

Fig. 5.35

the thickness of the whole symbol, according to whether the slope is in imaginery light or shade. One may assume the direction of light to be from above left (Fig. 5.34). This method, assisted by supplementary components, can be used to depict many small landforms (Fig. 5.35).

Further modifications can be introduced to distinguish the nature of the ground, e.g. grassy slopes, sand, scree or cliffs. A simple approach should be adopted and elaborate detailed symbols avoided, unless the precision of the survey demands it (Fig. 5.36).

5.3.3.6 THE PROBLEM OF AREA SYMBOLS

A recurring task for field scientists is to delimit the landscape according to a selected classification of vegetation, land use, geology, etc. While the reasons for selecting a particular classification are not the concern of this book, the means of mapping them are. The field plot of such a map will certainly have the boundary lines marked – only the surveyor will know if these lines represent the approximate or median location of a zone of change (as in soil mapping) or a precise line (as in a change in rock types along a fault). He may wish to differentiate between such levels of precision by using broken and continuous lines, but often, if the purpose of the map is to portray the areas of different quality, it is better if the lines are omitted, contrasts between area patterns themselves being sufficient (Fig. 5.37). A continuous line suggests a strong degree of continuity and precision and should be used only when the mapped data demand it.

The main problem might be to depict on the map, for example, ten different categories of vegetation or rock type, and possibly to permit the addition of plan and height information on the base map beneath. The first stage is to consult standard texts to discover if there are any existing conventions for the black and white depiction of such categories. Conventions, for example, exist in geology, although they are not rigidly maintained.

Fig. 5.37 Patterns in adjacent areas: the nature of boundary lines.

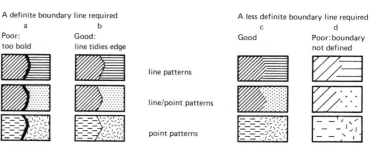

A definite boundary line required | A less definite boundary line required
a b | c d
Poor: too bold | Good: line tidies edge | Good | Poor: boundary not defined

line patterns

line/point patterns

point patterns

However, new requirements, eg 'bold boundary line' or 'vague boundary zone' could make columns 'a' and 'd' more suitable.

sandstone: regular grains
conglomerate: irregular particles
granite: crystalline structure
clay: fine bedding

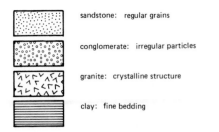

Fig. 5.38 Conventional symbols for solid geology.

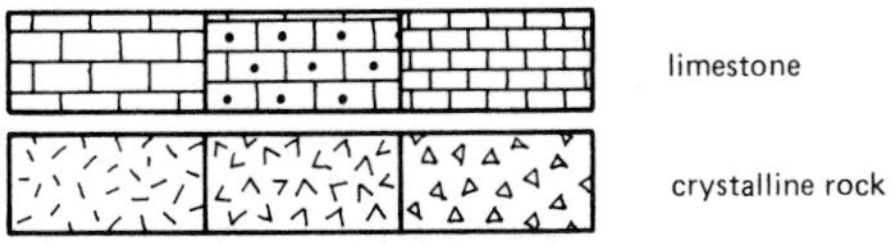

limestone
crystalline rock

Fig. 5.39 Some possible variations of geological map symbols.

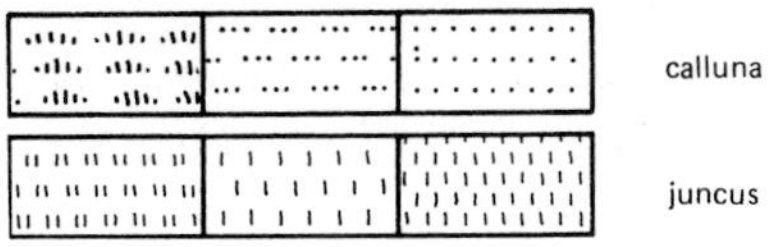

calluna
juncus

Fig. 5.40 Some possible variations of vegetation map symbols.

The symbols often employ connotation/imitation factors (Fig. 5.38). However, when a detailed map of rock facies is to be produced, these broad characteristics are insufficient and the symbols must be elaborated. Figure 5.39 depicts how the main design of the symbol for one rock type can be modified without losing the impression of its main characteristics. This principle can also be used for vegetation, etc. (Fig. 5.40). If a random method has been adopted, the map would be more difficult to read, owing to the 'noise' components of the various symbols.

In summary, if the map author requires maximum differentiation between each category, then maximum differentiation and contrast should be introduced into the area patterns. But if he wishes to retain the common features of broad classes, the design should reflect this also. In addition to these aspects of symbol design there are practical problems related to area symbols.

1. Size of area and size of pattern unit. Ensure that the spacing of lines, dots, etc., which make up the pattern is not so great that the pattern cannot be clearly identified (Fig. 5.41).
2. Different patterns have different levels of lightness value. Make sure that there are no instances of near-black and near-white patterns on the map, unless such contrast is vital to the design (Fig. 5.42).
3. Dimensions of the pattern elements. If the spacing and size of the dots or lines making up a pattern are too small it can create serious problems at the copying stage. Fine-grained area shading which looks impressive on the original drawing may appear blotchy or even disappear altogether after certain copying procedures.
4. Combining areas and base data. Clear differentiation may in some cases be easier to achieve by using visual levels – but by placing the

Fig. 5.41 Effect of symbol grain.

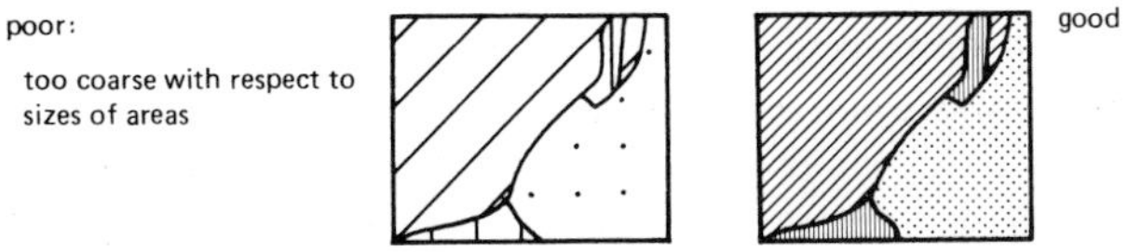

poor:
too coarse with respect to sizes of areas
good

Fig. 5.42 Area contrast.

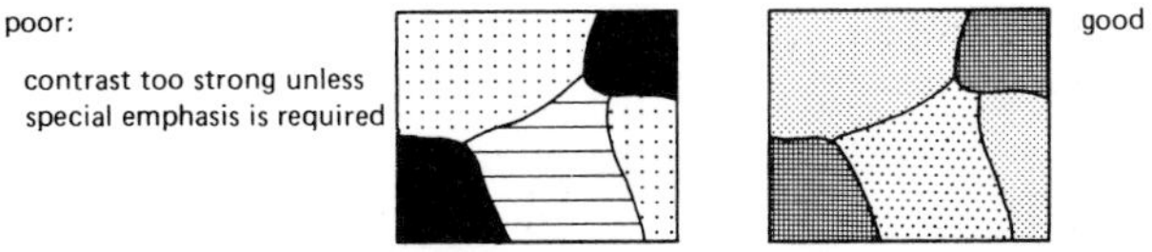

poor:
contrast too strong unless special emphasis is required
good

Fig. 5.43 Contrast between base map and theme.

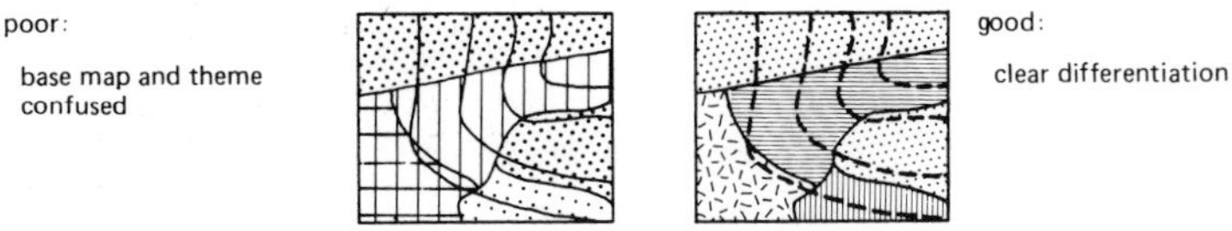

poor:
base map and theme confused
good:
clear differentiation

thematic area patterns in the 'background' with the clear lines of the base map in the 'foreground' (Fig. 5.43).

5. The use of colour. If only a few final copies are required, diazo prints could be made and the areas shaded in colour. The same logic as used with black and white patterns applies to colour, e.g. 'green' for grassland and 'brown' for arable, modified in tint and possibly having superimposed patterns. As before, existing conventions should be respected.

5.3.3.7 NAMES

Names or, more generally, text, can also be varied in graphic form, and these variables should be employed deliberately and not be permitted to happen by accident. The variables and their possible applications appear in Fig. 5.44, and Fig. 5.45 illustrates their application in a map.

A logical grouping of ground features and their symbols can be reflected by an appropriate selection of lettering style, e.g. physical features in serif style and cultural features in sans serif. However, accepting that some maps may later have to be copied using intermediate quality copiers, such as diazo, all the elements of the letters must be clear and strong. Beware of thin lines, especially in some seriffed letters, and of the enclosed parts of letters (counters) which may fill up and look solid if too small (Fig. 5.46).

In all the planning of this stage, it is necessary to take samples of parts of the map and to try out different designs and combinations. Names must also be positioned for accuracy and legibility (Fig. 5.47).

It is difficult to give precise rules for the

form:	CAPITALS lower case
	nature of feature or its importance
style/slope:	Serif sans serif
	UPRIGHT *ITALIC*
	type and classification of feature
width:	**condensed** extended
	size of feature or density of individual features
weight:	light **bold**
	importance of feature
size:	large small
	size and/or importance of feature
spacing:	t o o f a r a p a r t toodose
	adjusted to suit space available or to indicate good
	the extent of a feature Avoid extremes
orientation:	to suit map design – but maintaining horizontal if possible

Fig. 5.44 The variables of lettering.

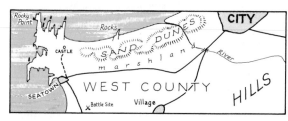

Fig. 5.45 Lettering variables on the map.

positioning of names, but certain circumstances should be avoided:

1. illegibility due to overlapping of names and symbols or even names and other names;
2. ambiguity of the item being named through poor

serif face

copy ABCDEFGHIJKLMNOPQRSTUVWXYZ

sanserif face

ABCDEFGHIJKLMNOPQRSTUVWXYZ

copy ABCDEFGHIJKLMNOPQRSTUVWXYZ
from the above

thin sections break up

ABCDEFGHIJKLMNOPQRSTUVWXYZ

good legibility preserved

Fig. 5.46 Letter quality suitable for copying.

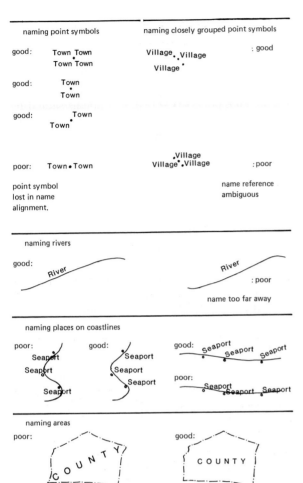

Fig. 5.47 Name arrangement.

Settlements:	large	**TOWN**
	medium	Village
	small	Farm
Physical Features:	large	*MOUNTAIN*
	medium	*Hill Tops*
	small	*Streams*

Fig. 5.48 Relative emphasis and the ranking of names for appropriate design.

location of the name; and

3. incomplete coverage, large features being represented by names that are too small.

Names should be located horizontally where possible, as this is the natural way to read text. However, if names must be reorientated it is best to curve them to suggest that they are either tending towards or away from the horizontal. On no account should names be inverted, or rendered with their letters in normal orientation but placed one above the other.

5.3.3.8 RELATIVE EMPHASIS

If the requirements demand different levels of importance to be placed on different items in the map, the contrast between all symbols and names can be adjusted to bring about the different levels of emphasis required. The map elements can be listed in order of importance, grouped if necessary, and the names listed alongside corresponding features. This ranking can then be translated into graphic form by modifying the symbols as described above (Fig. 5.48). The balance of marginal detail should also be considered in this way.

5.3.3.9 GRAPHIC EDITING

Once content has been selected, and symbols and names designed, adjustments may still have to be made to the positioning of names in particular, to ensure maximum legibility and to remove untidy overlapping and ambiguity (Fig. 5.49). At this stage, no final drawing will have been carried out, and such modifications can still be made.

5.3.3.10 LAYOUT

Accepting the requirements prescribed at the beginning – same-scale drawing, possibly for a smaller-scale reproduction – the frame must be placed around the map. Its proportions will be controlled mainly by the general shape of the region, which may be square or rectangular. However, if the map will eventually be incorporated into a publication, as a page or fold-out, this could restrict proportions: for example, a map may have to appear within A4, at a frame dimension of 26 cm × 16 cm, and this ratio will have to be scaled up to the drawing scale by means of the diagonal nomogram (Fig. 5.50).

During the survey the area may have been

Unedited Manuscript

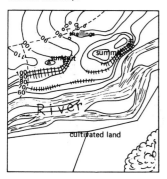

After graphic editing

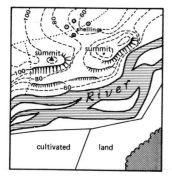

Fig. 5.49 Graphic editing.

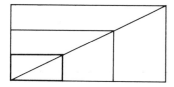

Fig. 5.50 Nomogram for scale change.

approached from the east, and thus the field plot might have been produced with north to the right. When it comes to final cartography, a decision must be made either to leave it as on the field plot or to re-orientate the image to the conventional 'north-at-the-top' form. The latter is desirable from the viewpoint of location diagrams, and placing the survey in a wider context through comparison with other maps, but the original layout might provide a more suitable framework. The choice should be governed by the specific case rather than by a general rule. Guidance on the nature and style of border grid values may be obtained from similar maps in standard topographic series.

Next, a complete checklist of marginal information should be prepared (Fig. 5.51), general proportions and lettering styles being added if desired. The sizes and styles will be dependent upon the size and character of the map, although the ranking provided is typical. Capitals should be used sparingly, in legend statements in particular, e.g.

'Unclassified Land', 'Unclassified land', or even 'unclassified land'; but *not*: 'The Survey Was Carried Out By . . .'.

Here, consistency is possibly more important than specific rules. There is no harm in consulting other similar maps of agreed high cartographic quality.

Small sketches (Fig. 5.52) can be used for planning the location of each item of marginal information. Suitable spaces around the map may be utilised, or a panel added at the bottom or down one side to contain the information. The key may require some extra consideration. It should remain compact, with items grouped logically. Avoid opening out the key merely to fill up space, since the result could break up the efficiency and continuity of the symbols therein (Fig. 5.53). The symbols must be identical to those on the map. There is an occasional tendency to make the key symbols similar in design but smaller in dimension. This is quite wrong! In the key panel the selected symbols may be placed to left or right (Fig. 5.54).

Occasionally, complex sequential combinations of symbols – for example estuaries, coastal zones and glacier fronts – may be represented best by model landscapes, where features can be seen in their correct settings (Fig. 5.55).

The design process has been analysed here in some detail. In many cases, especially with simple maps, the procedures detailed above can be carried out in a very short time. Finally, since proficiency in map design comes with experience, the mapmaker should always keep an open mind and be ready to make adjustments to design, even at the final drawing stage.

5.3.4 SPECIFICATIONS

Throughout the previous sections, the mapmaker should have been collecting notes and 'doodles' of

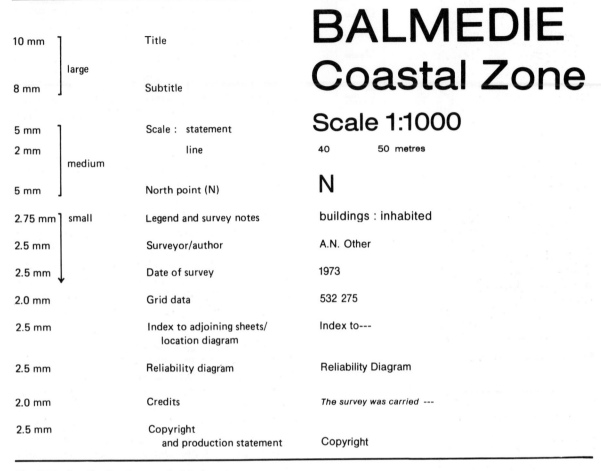

scale of type sizes Capitals height (example only)	subject	suggested styles
10 mm ⎤	Title	**BALMEDIE**
⎟ large		**Coastal Zone**
8 mm ⎦	Subtitle	
5 mm ⎤	Scale : statement	**Scale 1:1000**
2 mm ⎟	line	40 50 metres
⎟ medium		
5 mm ⎦	North point (N)	**N**
2.75 mm ⎤ small	Legend and survey notes	buildings : inhabited
2.5 mm ⎟	Surveyor/author	A.N. Other
2.5 mm ↓	Date of survey	1973
2.0 mm	Grid data	532 275
2.5 mm	Index to adjoining sheets/ location diagram	Index to---
2.5 mm	Reliability diagram	Reliability Diagram
2.0 mm	Credits	*The survey was carried ---*
2.5 mm	Copyright and production statement	Copyright

Fig. 5.51 Specification for marginal text.

Fig. 5.52 Layout.

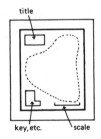
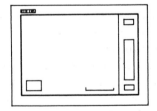
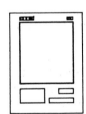

symbols, etc., which he feels are suitable. The specification is a list of the final design decisions reached, translated into practical terms and set out in a formal manner (Fig. 5.56). All items on the map should be included, grouped under point symbols, line symbols area symbols and text.

Construction
The design decisions do not necessarily indicate the exact method of execution, and it is convenient at

Fig. 5.53 The key panel.

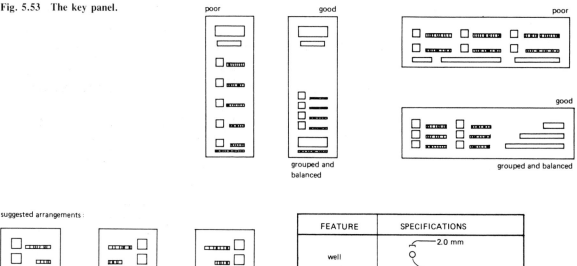

suggested arrangements:

Fig. 5.54 The key.

this stage to plan the approach to production in more detail (see Fig. 5.57).

In Fig. 5.57, Example A, the simplest case, step 2 will incorporate many of the design decisions. This and the layout should be completed in pencil before the final drawing is begun. If the final drawing is in ink only, and the layout and compilation have been completed, the act of drawing could almost be regarded as an exercise in tracing. In this case, in order to avoid untidy overlaps, etc., point symbols and names are produced first, followed by line symbols and patterns. If adhesive symbols and lettering are to be used, room should be left for them and they should be added after the ink drawing is complete, to save them from possible damage during the act of drawing (Fig. 5.58).

Figure 5.57, example B, incorporates some of the more sophisticated techniques of scribing and photography. The reader is referred to Keates

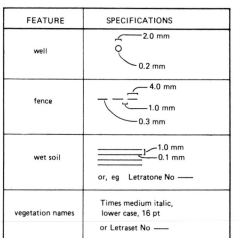

FEATURE	SPECIFICATIONS
well	⌐ 2.0 mm 0.2 mm
fence	⌐ 4.0 mm 1.0 mm 0.3 mm
wet soil	1.0 mm 0.1 mm or, eg Letratone No ——
vegetation names	Times medium italic, lower case, 16 pt or Letraset No ——

Fig. 5.56 Specifications.

(1989) for a more thorough treatment of the subject.

The type of flow diagram illustrated in Fig. 5.57 is worth producing, even for fairly simple maps, so that the plan is understood by the photogrammetrist, surveyor, cartographer and, if necessary, the author. It can also be expanded to produce a timetable and an estimate of the materials required.

When it comes to copying, it will be necessary to ensure that the final positive or negative reads the appropriate way (R, right reading; Я, wrong

Fig. 5.55

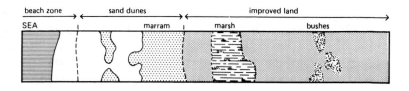

166

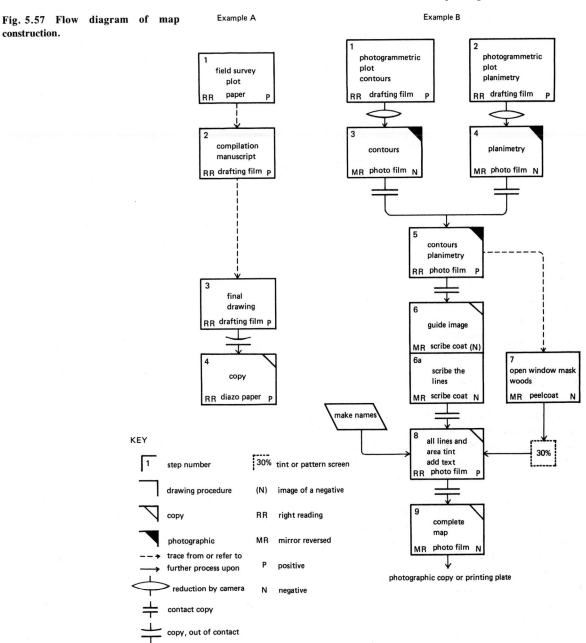

Fig. 5.57 Flow diagram of map construction.

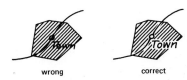

Fig. 5.58 Drawing order.

reading) on the image surface. For the production of a diazo print it is advisable to employ a wrong-reading positive to give maximum contact between positive and print, but for contact photographic prints a wrong-reading negative is normally required. If the project demands offset litho copies, advice on the whole production procedure should be sought from an experienced printer.

Computer assisted cartography

While standard techniques of generating x and y coordinates of mapped data are described elsewhere in this book, certain combinations of equipment and methods in both field surveying, and mapping from air photography or remotely sensed imagery, can produce these values directly. Also, conventionally plotted points, etc., can be converted to numerical or digital form through the use of a device called a digitiser. On this instrument points or lines are located or traced, and values are computed simultaneously. Once in digital form, and stored in a convenient data structure in computer compatible form, map data are then available for computer manipulation using appropriate software. Operations such as change-of-scale or-projection, contour interpolation and the generation of three-dimensional views are then relatively simple and fast. Hard copy of the results of these manipulations can be provided as listings in conventional alpha-numeric form, but graphic output is more common. The nature of this output will depend on the characteristics of the printing or plotting devices available, which vary in both image quality and in maximum available size. Considering the stated aims of this book, which assume fairly basic surveying techniques, equipment and experience on the part of the field scientist, these digital procedures are not elaborated further here. However, workers who may have access to suitable equipment and who may be inclined to use it, would be well advised to seek expert advice. Some useful references have been included in the bibliography.

Ultimately, the field scientist wants to know how to make or describe specifications for the production of a particular map or diagram. This part of the book provides some practical suggestions and indications of the more advanced possibilities, but does not attempt to treat the subject in its entirety.

FURTHER READING

Cuff, D. J. and Mattson, M. T., 1982, *Thematic Maps – Their Design and Production*. Methuen, London.

Dent, B. D., 1985, *Principles of Thematic Map Design*. Addison-Wesley, London.

Hodgkiss, A. G., 1970, *Maps for Books and Theses*. David and Charles, Newton Abbot.

I.C.A., 1984, *Basic Cartography for Students and Technicians*, Vol. 1., International Cartographic Association,

Imhof, Ed., *Cartographic Relief Presentation* (Edited by H. J. Steward). Walter de Gruyter, New York, 1982.

Keates, J. S., 1989, *Cartographic Design and Production*, 2nd Edition. Longman, Harlow.

Appendix 1 EXAMPLES OF ORDNANCE SURVEY INFORMATION

SELECTED LIST OF ORDNANCE SURVEY INFORMATION LEAFLETS (NEW SERIES)

No.
1   Index to Ordnance Survey Leaflets
2   History and work of the Ordnance Survey
11 Ordnance Survey maps and services for education
12 Large scale maps
30 Standard mapping services
31 Triangulation and minor control information
32 Survey information on microfilm and the supply of unpublished survey information
33 Levelling
44 Transparencies, map enlargements and reductions
47 Aerial photography
48 Digital mapping
49 Archaeological and historical maps
50 1: 50 000 Landranger Map Series of Great Britain
70 High- and low-water marks and tidal levels as shown on Ordnance Survey maps and Admiralty Charts
72 Facts and figures relating to the Transverse Mercator projection and National Grid
91 Metrication of Ordnance Survey maps
92 Index to 1: 250 000 Routemaster Series of Great Britain

HYPOTHETICAL BENCH MARK LIST

Description of bench mark	N.G. 10 m ref.	Altitude (Newlyn) (m)	Height of B.M. (above ground) (m)	Date of Levelling
NBM. RIVET ROCK, 9 m NE FROM RD NW SIDE TRK	8920 4127	18.721	–	11 55
NBM. RIVET ROCK, 31 m E FROM STR., 3 m N RD	8948 4120	8.893	–	11 55
FL. BRACKET S 8987 ON ANAIGH T.P.	8977 4100	26.711	–	05 62
RIVET ON ROCK, S SIDE TRACK, 8 m E FROM RD	8991 4105	24.937	0.60	05 62

All bench marks in this list fall on 1:2500 sheet NI8941.

SAMPLE TRIANGULATION STATION INFORMATION SHEET

Triangulation Station Information

Name of Station	GREENHILL
Station No.	NX/03/T11
Type of Mark	PILLAR
Order in Trig Network	3RD
Nature of Ground	PASTURE
Route and Access	SEE OWNER
KM Number	NX9520
1 in Sht No.	272 (7th Series)
F1 Br No	S7230
County	RAMPSHIRE
Date of Coordinates	1953

Name, Addresses of Owner,   General Aspect
Agent and Tenant   & Obstructions
Owner MR W. SMITH,
        GREENHILL,
        WHITECAIRNS,
        RAMPSHIRE.
N.G. Ref.
Agent
N.G. Ref.   Good Visibility
Genant   in All Directions
N.G. Ref.

Name of station: GREENHILL   Station No. NX/03/T11

Description
(Measurements in metres)

No description   Station on highest part of hill

Bearings to	km Sq	Dist Approx. (kms)	degrees	mins.	sec.
Kincraig Chy	NX9624	4.8	12	22	50
Foveran (Pillar)	NX9923	5.6	53	50	29
Smith House Spire	NX9618	2.4	134	49	41
Beauty Hill (Pillar)	NX9020	4.3	276	17	34

National Grid Re-Triangulation Coordinates

Pillar E	395 075,28	N	920 052.32
E		N	
E		N	
E		N	
E		N	

Altitude on Newlyn Datum to FI Br 374.34 ft 114.10 m

The information on this Triangulation Station with the exception of the coordinates, altitude and grid bearings was completely revised in 1969.

Both types of information, i.e. height and planimetry, can be obtained from ordnance Survey, Romsey Road, Maybush, Southampton SO9 4DH

Appendix 2 SOURCES OF AERIAL PHOTOGRAPHY AND EARTH SATELLITE IMAGERY

CENTRAL REGISTERS OF AERIAL PHOTOGRAPHY

1. For Scotland
   Scottish Development Department, Air Photographs Unit, New St. Andrews House (Room 1/21), Edinburgh EH1 3SZ.
2. For Wales
   Welsh Office, Air Photographs Unit, Crown Offices, Cathay Park, Cardiff CF1 3NQ.
3. For the British Isles
   The Cambridge Committee for Aerial Photography, the Mond Building, Free School Lane, Cambridge CB2 3RF.

The Central Register for England, previously housed at the Department of the Environment, is now closed. The most comprehensive single source of aerial photography is now the Ordnance Survey, Romsey Road, Maybush, Southampton S09 4DH.

PRINCIPAL AERIAL SURVEY COMPANIES

BKS Surveys Ltd., 47 Ballycairn Road, Coleraine, Co. Londonderry, Northern Ireland BT51 3HZ.
Clyde Surveys Ltd., Reform Road, Maidenhead, Berkshire SL6 8BU.
Hunting Surveys Ltd., Elstree Way, Boreham Wood, Herts. WD6 1SB.
Meridian Airmaps Ltd., Marlborough Road, Lancing, Sussex BN15 8TT.
J. A. Storey and Partners, 92–94 Church Road, Mitcham, Surrey CR4 3TD

NON-BRITISH SOURCES OF AERIAL PHOTOGRAPHY

Canada
National Air Photo Library, 615 Booth Street, Ottawa, Ontario, Canada K1A OE9.

United States of America (a selection)
User Services Unit, EROS Data Centre, Sioux Falls, South Dakota 57198, USA.
Agricultural Stabilisation and Conservation Service and US Dept. of Agriculture, Aerial Photography Field Office, 2505 Parley's Way, Salt Lake City, Utah 84109, USA.
Coastal Mapping Division, C3415, National Ocean Survey, NOAA, Rockville, Maryland 20852, USA.
For a complete list of North American sources see D. Kroek, 1979, in Bibliography.

Since most modern topographic mapping is done from aerial photography, the agencies already listed as map sources will have air photo archives.

SOURCES OF EARTH SATELLITE DATA AND IMAGERY

UNITED KINGDOM
National Remote Sensing Centre, Space Dept., R 190 building, Royal Aircraft Establishment, Farnborough, Hampshire GU14 6TD. (Also provides information on other parts of the world).

Europe
Earthnet User Services, C.P. 64, 00044 Frascati, Italy. The following countries maintain a National Point of Contact with Earthnet for priority access to and distribution of data of their national territory: Austria, Belgium, Denmark, France, Federal Republic of Germany, Ireland, Italy, Netherlands, Norway, Spain, Sweden, Switzerland and the United Kingdom.

Canada
Canada Centre for Remote Sensing, 2464 Sheffield Road, Ottawa, Ontario, Canada K1A OY7.

United States of America
User Services Unit, EROS Data Centre, Sioux Falls, South Dakota, 57198, USA.

Appendix 2   Sources of aerial photography and earth satellite imagery

Technology Application Centre, University of New Mexico, Code 11, Albuquerque, New Mexico 87131, USA.

Earth Observation Satellite Company (EOSAT), 4300 Forbes Blvd., Lanham, Maryland 20706, USA.

BIBLIOGRAPHY

Anderson, J. R., Hardy, E. E., Roach, J. T. and Witmer, R. E., 1976. A land use and land cover classification system for use with remote sensor data. Professional Paper 964. US Geological Survey.

Avery, T. E. and Berlin, G. L., 1985, *Interpretation of Aerial Photographs* (4th edn). Burgess, Minneapolis, USA.

Colwell, R. N., (ed.) 1960, *Manual of Photographic Interpretation*. American Society of Photogrammetry, Falls Church, Virginia, USA.

Dickinson, G. C., 1979. *Maps and Air Photographs* (2nd edn.) Edward Arnold, London.

Ford, J. P., Cimino, J. B. and Elachi, C., 1983. *Space Shuttle Columbia Views the World with Imaging Radar: The SIR-A experiment*, Publication 82-95. Jet Propulsion Laboratory, Pasadena, California.

GEOCENTER, 1986 and annually, *GEO Katalog 1 and 2*. Geocenter, International Map Centre, Stuttgart, Federal Republic of Germany.

Howard, J. A., 1970, *Aerial Photo-ecology*. Faber and Faber, London.

Keates, J. S., 1972, Symbols and Meaning in Topographic Maps, *International Yearbook of Cartography* 12.

Keates, J. S., 1989, *Cartographic Design and Production.* 2nd Edition. Longman, London.

Keates, J. S., 1982, *Understanding Maps*. Longman, London.

Kroeck, D., 1979, *Everyone's Space Handbook – A Photo-imagery Source Manual.* Arcata, Pilot Rock, California.

Lewis, A. J. (ed.), 1976, *Geoscience Applications of Imaging Radar Systems. Remote Sensing of the Electromagnetic Spectrum* (Special edition of *Annals of the Association of American Geographers*, **3**, no. 3.

Lo, C. P. 1986 *Applied Remote Sensing.* Longman, London.

Lo, C. P. and Wong, F. Y., 1973, Micro-scale geomorphology features, *Photogrammetric Engineering*, **39**, no. 12, pp. 1289–96.

Malan, O. G., 1981. *How to Use Transparent DIAZO Colour Film for Interpretation of Landsat Images.* COSPAR Technique Manual Series, no. 9, 33 pp. COSPAR Secretariat, 51 bd de Montmorency, 75016 Paris.

McCoy, R. M., 1969, Drainage network analysis with K-band radar, *Geographical Review*, **59**, 493–512.

Methley, B. D. F., 1970, Heights from parallax bar and computer, *The Photogrammetric Record*, **6**, no. 35, 459–65.

Monmonier, M. A. 1982 *Computer-Assisted Cartography: Principles and Prospects.* Prentice Hall, Englewood Cliffs.

Nüesch, D. R., 1982, *Augmentation of Landsat MSS data by Seasat SAR imagery for agricultural inventories* (Remote Sensing Series, Vol. 7). Department of Geography, University of Zurich.

Paine, D. P., 1981, *Aerial Photography and Image Interpretation for Resource Management.* Wiley & Sons, New York.

Parry, R. B. and Perkins, C. R., 1987 *World Mapping Today*. Butterworths, Sevenoaks.

Remote Sensing Society, 1983, Submission to the House of Lords Select Committee on Science and

Technology (Sub-committee 1: Remote Sensing and Digital Cartography).

Rhind, D. W. and Adams, T., 1982 Computers in Cartography. Special Publication No. 2, British Cartographic Society, London.

Rhind, D. and Hudson, R., 1980, *Land Use*. Methuen, London.

Ryerson, R. A. and Gierman, D. M., 1975. A remote sensing compatible land use activity classification, Technical Note 75–1. Canada Centre for Remote Sensing.

Schwidefsky, K., 1961, *Outline of Photogrammetry* (translated from the German by J. Fosberry). Pitman, New York.

Slama, C. C. (ed.), 1980, *Manual of Photogrammetry* (4th edn). American Society of Photogrammetry, Falls Church, Virginia.

Tait, D. A., 1970, Photo-interpretation and topographic mapping, *The Photogrammetric Record*, **6**, no. 35, 466–79.

Thompson, E. H., 1954, Heights from parallax measurements, *The Photogrammetric Record*, **1**, no. 4, 38–49.

Ulaby, F. T., Moore, R. K. and Fung, A. K., 1981, *Microwave Remote Sensing: Active and Passive* **1** *Fundamentals and Radiometry*. Addison-Wesley, Massachussetts.

Welch, R. and Marko, W., 1981, Cartographic potential of a spacecraft line array camera system: Stereosat, *Photogrammetric Engineering and Remote Sensing*, **47**, no. 8, 1173–85.

Winch, K. L. (ed.), 1976, *International Maps and Atlases in Print* (2nd edn). Bowker, London.

Wright. R. and Hubbard, N. K., 1982, Some problems associated with large area mapping from Landsat, *Proceedings of International Society for Photogrammetry and Remote Sensing*, Commission IV Symposium, Crystal City, Virginia, 263–71.

Yoeli, P. 1982 Cartographic Drawing with Computers. Computer Applications, Special Issue, Volume 8. Department of Geography, University of Nottingham.

INDEX